I0031274

Smart Grid Dictionary®

Christine Hertzog

6th Edition

Smart Grid Dictionary

Copyright © 2014 Christine Hertzog

Booksite: www.smartgridlibrary.com

All rights reserved under International and Pan-American Copyright conventions, including the right to reproduce this book or portions thereof in any form whatsoever.

ISBN number: 978-0-9898893-1-5
6th Edition – October 2014

Doug Stuart, cover and book designer

To submit suggestions for new terms and acronyms, visit www.smartgridlibrary.com

Published by GreenSpring Marketing LLC

Table of Contents

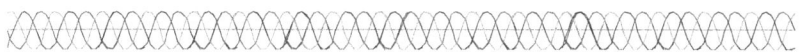

Acknowledgements

I am grateful for the generous and invaluable contributions of time, knowledge, and advice from the Smart Grid Dictionary Advisory Board, whose members suggested new terms, acronyms, and jargon; and contributed and reviewed definitions to help ensure the overall quality and accuracy of the contents.

The Smart Grid Dictionary Advisory Board members are: Katherine Hamilton, Founder at 38 North Solutions; Terry Mohn, Chief Executive Officer for General MicroGrids; Paul Myrda, Technical Executive at the Electric Power Research Institute (EPRI); Paul De Martini, Managing Director, Newport Consulting Group; Jeanne Boyce, Manager, Innovation Management at Southern California Edison, Advanced Technology at Southern California Edison; Clint Cooper, Co-founder of Teocalli Partners LLC; Mike Ahmadi, Vice President of Operations at GraniteKey LLC; Christie Jordan, Executive Director of the Alliance for Water Education; and Rebecca Herold, CEO, The Privacy Professor and Partner, Compliance Helper.

These industry professionals have technical and policy expertise and are well known and respected for their contributions to building and advancing Smart Grid knowledge and benefits. Thank you for your exceptional help with content for the Smart Grid Dictionary 6th Edition.

A special thanks to Justin Hagler for his invaluable Dictionary content research.

Thanks to all my readers, colleagues, and friends who give me new terms, advice, and encouragement for this Edition.

About the Author

Christine Hertzog is the founder and Managing Director of the Smart Grid Library and SGL Partners, delivering consulting and information services about Smart Grid and Smart Infrastructure technologies, services, and solutions. Her firm provides pragmatic guidance to vendors, governmental entities, and utilities covering a broad range of needs such as strategic corporate and market insights and design and deployment of prosumer-centric utility operations.

Ms. Hertzog is the author of the Smart Grid Dictionary that defines the jargon, acronyms, and terminology about technologies, international standards, and organizations associated with the Smart Grid and Smart Infrastructure. She is the co-author of The Smart Grid Consumer Focus Strategy, which identifies consumer/utility challenges and methods to ensure successful prosumer operations and interactions, and a soon to be published book on the Smart Grid and data privacy. She is a recognized thought leader and regular speaker at industry conferences and writes a syndicated blog about Smart Grid and Smart Infrastructure topics.

Based in Silicon Valley, Ms. Hertzog serves as an advisor to Smart Grid startups, as well as industry associations and publications including The Energy Collective, ElectricityPolicy.com, Energy Post, and IBCon. She has a Master of Science degree in Telecommunications from the University of Colorado – Boulder.

Introduction

Welcome to the Smart Grid Dictionary, 6th Edition. This Dictionary is a valuable resource for all Smart Grid, Smart Infrastructure, and M2M communications stakeholders. Whether you work for a utility, regulatory or other policy agency, vendor, or have an interest in the Smart Grid, Smart Infrastructure, and M2M business sectors, this Dictionary will help you understand its terminology, standards, regulatory agencies, R&D entities, industry associations, and technology trends.

This edition contains almost 2400 terms and acronyms covering generation, transmission, distribution, storage, and consumption of electricity; transmission, distribution and consumption of water; network and cyber security; sensor technologies; energy efficiency; building automation; standards; and telecommunications. Selected natural gas terminology has been added since it uses Smart Grid technologies for monitoring, measuring, and managing transmission, distribution, and consumption. This Edition also adds relevant M2M terminology that applies to Smart Grids, intelligent buildings, and smart cities. It also includes data privacy terms pertinent to energy usage data concepts, policies, and organizations.

Definitions were researched and developed from a variety of sources including websites for DOE (Department of Energy); NIST (National Institute of Standards and Technology); industry and trade associations; standards organizations; industry-specific and general media websites; and content derived from conferences, articles, webinars, seminars, presentations, books, and whitepapers. The Smart Grid Dictionary Advisory Board (see Acknowledgements) also provided invaluable contributions of new terms and definition content.

This Dictionary organizes definitions by acronyms, since websites, articles, and presentations do not spell out acronyms. For instance, if you are looking for the Department of Energy definition, look under DOE, its acronym.

Numbers

100-hour coincident peak
Average of demands during the highest 100 load hours for a utility. It is one type of measure of peak capacity that is used in regulatory cost of service models.

1xRTT (One Times Radio Transmission Technology)
A term that refers to an operating mode of CDMA2000 in cellular networks. The numeral 1 refers to one 1.25-MHz channel. It provides high-speed data services and voice. It is technically considered a 3G system, but is sometimes classified as "second and a half" generation because it is slower than 3G. 1xRTT is rated at 144K upload and download maximums, but the typical speeds are 50K to 80K.

2G
A term for the second generation of technologies in mobile carrier networks. The primary standards are TDMA (Time Division Multiple Access) and CDMA (Code Division Multiple Access) with GSM prevalent in most of the world, and iDEN, a proprietary standard found in North America. CDMA standards are branded cdmaOne™ and include IS-95A and IS-95B. From an M2M perspective, 2G does not support IPv4 or IPv6 addressing.

3DES (Triple Data Encryption Standard)
A NIST (National Institute of Standards and Technology) standard using a data encryption algorithm (DES) that is applied three times to each data block. It enables better protection than DES and allows continued use of those algorithms.

3G
A term that describes the third generation in mobile systems evolution. 3G systems have higher data transmission speeds and always-on data access

and typically use new radio spectrums that were not available to 2G networks. The most prevalent technology standard for 3G is UMTS. From an M2M perspective, 3G does not support IPv4 or IPv6 addressing.

3GPP (3rd Generation Partnership Project)
A project of global telecommunications organizations to create 3G and mobile broadband specifications based on GSM. The standards are used by cellular phone networks, and their primary function is to outline network protocols for 2G, 3G, and 4G packet delivery. The partners in this project are European Telecommunications Standards Institute, Association of Radio Industries and Businesses/Telecommunication Technology Committee, China Communications Standards Association, Alliance for Telecommunications Industry Solutions, and Telecommunications Technology Association. (http://www.3gpp.org/)

3GPP2 (3rd Generation Partnership Project 2)
A project of global telecommunications organizations to create 3G specifications based on CDMA. The five collaborating SDOs (Standards Development Organizations) are Association of Radio Industries and Businesses, Communications Standards Association, Telecommunications Industry Association, Telecommunications Association, and Telecommunication Technology Committee.

4G
A term used to describe the next generation of wireless, broadband mobile communications that differs from 3G technologies in the following factors: higher speeds ranging from 100MB for highly mobile use to 1GB for local Wi-Fi and nomadic use; automatic roaming across multiple networks for seamless handoffs; and support for voice, data, video, M2M (machine to machine), and location-aware applications. It may be based on OFDM (Orthogonal Frequency Division Multiplexing). The ITU (International Telecommunications Union) is working on a formal definition. From an M2M perspective, 4G does support IPv4 or IPv6 addressing.

6LoWPAN (Ipv6 Low power Wireless Personal Area Network)
The name of a protocol and an IETF (Internet Engineering Task Force)
working group with the objective to define adaptations between IPv6 and
IEEE 802.15.4 networks. It is characterized by extreme low power to run
potentially for years on batteries and extreme low cost. One result could be
interoperability between low-power devices and existing IP devices with
standard routing techniques.
(http://datatracker.ietf.org/wg/6lowpan/charter/)

A

A4WP (Alliance For Wireless Power)
An independently operated, nonprofit organization whose members include electronics industry leaders focuses on wireless technologies. The goal of the organization is to assist and encourage the development of integrated wireless power transfer technology that will eliminate the need to physically plug in mobile electronic devices in order to charge their batteries. (www.a4wp.org)

AAA (Authentication, Authorization, and Accounting)
A protection, validation, and usage tracking infrastructure used in network access security. AAA servers are very important to CDMA data networks, currently the primary communication protocols for V2G.

AB32 (Assembly Bill 32 – Refers to the California Global Warming Solutions Act of 2006)
Landmark legislation in the state of California that sets an economy-wide cap on California greenhouse gas emissions at 1990 levels by no later than 2020.

AB118 (Assembly Bill 118 – Refers to the Alternative and Renewable Fuel and Vehicle Technology Program)
A California program created in 2007 with the passage of AB118 to increase the use of alternative and renewable fuels and innovative technologies that will transform California's fuel and vehicle types to help attain the state's climate change policies. It is managed by the CEC (California Energy Commission).

AB1150 (Assembly Bill 1150 – Self-Generation Incentive Program)
A California law enacted in 2012 that authorizes the California Public Utilities Commission to collect funds for the Self-Generation Incentive Program (SGIP) through December 31, 2014. It extends a program that

encourages deployment of distributed generation that is integrated into the electrical grid to improve efficiency and reliability.

AB2514 (Assembly Bill 2514 – California Energy Storage Bill)
A California law enacted in 2010 that requires the state PUC (Public Utilities Commission) to open a proceeding by March 2012 to consider establishing IOU(investor owned utility) procurement targets for energy storage systems to be achieved by 2015, with an additional target by 2020. Publicly owned utilities have comparable requirements for energy storage and shifting peak energy usage.

ABPA (American Backflow Prevention Association)
An organization focused on protecting drinking water from contamination through cross-connections. (http://www.abpa.org/introduction.htm)

AC (Alternating Current)
Electric current that reverses its direction at regularly recurring intervals.

ACAR (Aluminum Cable, Alloy-Reinforced)
A transmission line that consists of a stranded cable made with an aluminum alloy that has low electrical resistance and good strength. Typically used for high-voltage lines.

Access BPL (Access Broadband PowerLine)
The use of powerline communications outside the home.

ACE (Area Control Error)
The difference between scheduled and actual electrical generation within a control area on the power grid. This calculation incorporates the effects of frequency bias and correction for meter errors and is used by Balancing Authorities to comply with NERC (North American Electric Reliability Corporation) reliability standards. It is playing an increasing role as more intermittent renewable energy sources are included in grid operations.

ACEEE (American Council for an Energy-Efficient Economy)
A nonprofit organization dedicated to advancing energy efficiency as a means of promoting economic prosperity, energy security, and environmental protection. ACEEE's program areas include federal and state energy policy, research regarding various aspects of energy efficiency, and communications to energy efficiency resources. (http://www.aceee.org/)

Accelerometers
A type of sensor that detects motion.

ACER (Agency for the Cooperation of Energy Regulators)
An organization formed in early 2011 that complements and coordinates the work of the national regulatory agencies (NRAs). It is intended to help create European network rule; decide terms and conditions for access and operational security for cross border infrastructure; give advice on energy-related issues to the European institutions; and report to the European Parliament and CEER. (http://www.energy-regulators.eu/portal/EER_HOME/EER_ABOUT/Tab)

ACORE (American Council on Renewable Energy)
A nonprofit organization with members from every aspect and sector of the renewable energy industries and their trade associations, including wind, solar, geothermal, biomass and biofuels, hydropower tidal/current energy and waste energy. The scope of membership within ACORE also includes financial institutions, government leaders, educators, end users, professional service providers and allied nonprofit groups. (http://www.acore.org/front)

Acre-foot
A common water metric that describes the water volume to cover one acre of land with a one-foot depth of water. Equivalent to 325,851 gallons or 1,230 liters.

ACSI (Abstract Communication Services Interface)
A part of IEC 61850 that defines communications between a client and a remote server and communications of one device to many devices for system-wide event distribution, as well as resolution of tracking and security issues. Also plays a role in IoT communications for intelligent electronic devices.

ACSR (Aluminum Cable, Steel-Reinforced)
A transmission line that consists of a stranded steel core to bear the mechanical load with layers of stranded aluminum around this steel core to transmit the current.

Active power
The electric power that performs work, measured in kilowatts or megawatts. Also known as real or true power.

Actual peak reduction
The actual reduction in annual MW (megawatt) peak load achieved through participation in demand response programs at a utility or an RTO/ISO level during annual system peak loads. From a CSP (Curtailment Service Provider) perspective, it is the demand response load provided during the peak for the region or territory of their aggregated customer load.

Actuator
A device that converts an electric or pneumatic signal into an action. Types of actuators include electronic, electric, solenoid, pneumatic, thermostatic radiator and electrohydraulic.

ADA (Advanced Distribution Automation)
A collection of intelligent sensors, remote controllers, and bi-directional communications to manage distribution grids — covering substations to AMI assets.

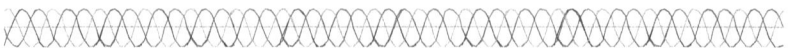

ADDRESS (Active Distribution network with full integration of Demand and distributed energy RESourceS)
A four-year project (2008 – 2012) co-founded by the European Commission under the 7th Framework Programme (FP7) and carried out by a consortium of 25 partners from 11 European countries. Its purpose is to enable the active participation of small and commercial consumers in power system markets and provision of services to the different power system participants. (http://addressfp7.org/)

ADE (Automated Data Exchange)
An M2M (machine to machine) exchange of data that is enabled by a CIM (Common Information Model) that serves as a standardized or universal data model so that different systems can exchange information with reduced data conversion routines. This term also refers to any system that enables access to customer meter data for either customer or authorized third-party use.

Adequacy
One of two primary characteristics of a reliable interconnected bulk power system. It is the ability of the bulk power system to supply the aggregate electrical demand and energy requirements of the customers at all times, taking into account scheduled and reasonably expected unscheduled outages of system elements.

ADI (ACE Diversity Interchange)
The process of pooling individual Area Control Errors or ACE to take advantage of control error diversity to reduce control burdens and accommodate integration of more intermittent renewable energy sources into the grid.

Adjusted electricity
An electricity measurement that includes the approximate amount of energy used to generate electricity.

ADS (Association for Demand response and Smart grid)
A nonprofit organization formed in 2004 to increase the knowledge base in the USA on demand response and facilitate the exchange of information and expertise among policymakers, utilities, system operators, technology companies, consumers, and other stakeholders. Members include utilities, ISOs (Independent System Operators), technology vendors, and retailers. Previously known as DRCC (Demand Response Coordinating Committee). (www.demandresponsesmartgrid.org/)

Advanced Analytics
Software solutions that can manage the increased data resulting from Smart Grid technologies, ranging from distribution automation, WASA (Wide Area Situational Awareness)), and smart meters, and transform it into manageable information for decision-making, planning, and forecasting.

Advanced biofuel
As defined in the ARRA Section 1306 amending EISA 2007 (Energy Independence and Security Act of 2007), any renewable fuel that meets a 50-percent life-cycle GHG emissions reduction from the petroleum baseline and is not derived from corn starch is considered to conform to EISA 2007 USA renewable fuel standards. Fuel sources can include biomass or vegetable and algae oils.

Advanced field telecommunications
Utility communication networks embedded in transmission and distribution facilities that deploy IEDs (Intelligent Electronic Devices) for remote monitoring and management to improve responsiveness to changing conditions.

Advanced grid materials
Composites such as nanomaterials or other metal and mineral fabrications that are incorporated into electrical generation, transmission, or

distribution components ranging from advanced conductors to energy-storage technologies to improve the capacity, performance, and reliability of small- to large-scale grids.

Advanced load control
The practice of using programs that reduce electricity use in residential, commercial, and/or industrial customers, such as automated HVAC controls, lighting management, and smart appliance interactions to address peak electricity demand conditions instead of bringing online additional generation facilities.

Advanced meter
See Smart Meter.

Advanced Volt-VAR Control
See AVVC.

AEDG (Advanced Energy Design Guides)
A series of documents targeted to specific building types such as commercial office buildings or K-12 schools that deliver energy efficiency recommendations for energy savings of at least 30 percent in any climate. (http://www.ashrae.org/publications/page/1604)

AEE (Association of Energy Engineers)
A nonprofit and global professional society working to promote the scientific and educational interests of those engaged in the energy industry and to foster action for sustainable development. Members include the commercial, industrial, institutional, governmental, energy services, and utility sectors. (http://www.aeecenter.org/)

AEIC (American Energy Innovation Council)
An industry association focused on developing new energy technologies to build economic growth, create jobs in new industries, and reestablish America's energy technology leadership through public investments.

AEIC (Association of Edison Illuminating Companies)
A global electric industry association seeking improvements to the technological aspects of the industry through six committees focused on the following: power generation, power delivery, electric power apparatus, load research, meter and service, and cable engineering. Members include electric utilities, generating companies, transmitting companies, and distributing companies (investor-owned, federal, state, cooperative, and municipal systems.) (www.aeic.org/)

AEMC (Australian Energy Market Commission)
A national and independent rule maker and developer for Australian energy markets – electricity and natural gas. It also provides strategic and operational advice to the Council of Australian Governments' Ministerial Council on Energy to ensure efficient, reliable and secure wholesale and retail energy market frameworks for consumers. (http://www.aemc.gov.au)

AEO (Annual Energy Outlook)
Yearly report produced by the EIA (Energy Information Administration) that forecasts electricity consumption.

AER (Australian Energy Regulator)
Australia's national energy market regulator and an independent statutory authority. The AER operates under the Competition and Consumer Act 2010 and mostly relates to energy markets in eastern and southern Australia. Functions include setting usage charges (electricity and gas) to transport energy to customers; monitoring and enforcing compliance in wholesale electricity and gas markets; and publishing information on energy markets. (www.aer.gov.au)

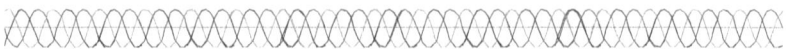

Aeration
Water treatment processes that mix air and water to dissipate contaminants.

AES (Advanced Encryption Standard)
A NIST (National Institute of Standards and Technology) standard for data encryption that uses algorithms based on iterative, symmetric key block ciphers. It supersedes DES as the encryption standard. It is also known as Rijndael and FIPS (Federal Information Processing Standard) 197, and is included in the ISO-IEC 18033-3 standard.

AESO (Alberta Electric System Operator)
A not-for-profit ISO organization responsible for the safe, reliable, and economic planning and operation of the AIES (Alberta Interconnected Electric System). It develops and administers transmission tariffs, procures ancillary services to ensure system reliability, and manages settlement of the hourly wholesale market and transmission system services, among other duties. It is one of the 10 ISO/RTOs (Independent System Operator/Regional Transmission Organization) currently operating in North America. (www.aeso.ca)

AESP (Association of Energy Services Professionals)
A not-for-profit association focused on improving the delivery and implementation of energy efficiency, energy management, and distributed renewable resources. Members include IOUs (Investor Owned Utilities), municipal and cooperative utilities, and manufacturers. (www.aesp.org)

AFUDC (Allowance for Funds Used During Construction)
An accounting term that is also known as interest during construction.

AFV (Alternative Fuel Vehicle)
Any vehicle that does not use petroleum-based fuel. Federal agencies are mandated by EPACT (the Energy Policy Act), Executive Order 13423, and

EISA (the Energy Independence and Security Act) to purchase alternative fuel vehicles, increase consumption of alternative fuels, and to reduce petroleum consumption. (http://www.fueleconomy.gov/feg/current.shtml)

AFV OIR (Alternative Fuel Vehicle Order Instituting Rulemaking)
A California PUC action to evaluate policies to develop infrastructure to overcome barriers to the widespread deployment and use of PHEVs (plug-in hybrid electric vehicles) and EVs (electric vehicles).

Aggressive water
Water that is acidic or contains corrosive substances that can damage pumps, plumbing, piping and devices.

AGWT (American Ground Water Trust)
A nonprofit organization formed to protect ground water resources and educate the public in water resource decisions and management through educational materials, workshops, and conferences. http://www.agwt.org/

Alternative fuel
Section 301 of EPACT (the Energy Policy Act) defines alternative fuels as biodiesel, denatured alcohol, electricity, hydrogen, methanol, mixtures containing up to 85% methanol or denatured ethanol, natural gas or propane. This definition was expanded by the National Defense Authorization Act of 2008 to include any other type of vehicle that reduces petroleum consumption including fuel cell vehicles, hybrid electric vehicles, and advanced lean burn technology vehicles.

AGC (Automatic Generation Control)
The main wide-area control system that controls tie-line power flows and generator outputs. It is part of the online feedback controls to generators, typically via the Internet, and helps system operators deliver regulation services.

Ah (Ampere hour)
A unit of electrical charge or current, commonly used in battery ratings to determine the length of time for discharge at a desired charge level.

AHAM (Association of Home Appliance Manufacturers)
A not-for-profit industry association representing manufacturers of major and portable home appliances, floor care appliances, and their suppliers. It develops voluntary technical standards and submits some of these to ANSI (American National Standards Institute) for approval. (http://www.aham.org/)

AIM (Association for Automatic Identification and Mobility)
A global trade association representing automatic identification and mobility technology vendors in the promotion of emerging technologies and standards for bar codes, RFID (Radio Frequency IDentification) and mobile computing. It works with ANSI and ISO in standards development and adoption. (http://www.aimglobal.org/)

Air conditioning intensity
The ratio of AC consumption to cooled square footage and cooling degree-days (the baseline is 65 degrees F). This measurement is a way to compare different types of housing units and households by controlling for differences in housing size and weather.

ALCA (Advanced Load Control Alliance)
An industry association that promotes load control for utility deployments in peak-demand reduction, economic dispatch, and/or T&D reliability. It shares information on marketing, deployments, benefit costs, and technology attributes gathered from participating members and from information available in the public domain. (http://www.alca.info/)

AllJoyn

An open-source application development framework developed by Qualcomm for proximity-based M2M communications. It is based on peer-to-peer mesh networking to automatically discover and connect devices using a number of wireless protocols. It is now managed by the Linux Foundation. (www.alljoyn.org)

AllSeen Alliance

An industry organization focused on the development of a universal framework for IoT applications based on AllJoyn, an open source application development framework designed by Qualcomm but now managed by the Linux Foundation. (www.allseenalliance.org)

AM (Amplitude modulation)

A technique used to transmitting information via radio carrier wave in which the voltage applied to the carrier is varied over time.

American Biogas Council

An industry association focused on policy advocacy and education about biogas, anaerobic digestion, and renewable energy. (http://www.americanbiogascouncil.org)

American Energy Innovation Council

The mission of this group of business leaders is to encourage the development of new energy technologies. (http://www.americanenergyinnovation.org/)

American Planning Association

A not-for-profit educational organization that provides leadership in vital community development. Its primary function is to serve as a forum for city and regional planning in the United States, with a focus on sustainability and serving public interests. (www.planning.org)

AMI (Advanced Metering Infrastructure)
Electricity meters, bi-directional communications network hardware and software, and associated system and data management software that measures and records usage data at set intervals, and provides usage data to consumers, utilities, and other parties at set intervals. The set intervals are specified by regulatory agencies.

AMI-Ent TF (AMI Enterprise Task Force)
A sub-group within the UtilityAMI working group of the Open Smart Grid Subcommittee in the UCAIug. It defines requirements, policies, and services, based on utility industry standards such as the CIM (Common Information Model), required for information exchange and control between the MDMS (Meter Data Management System) and enterprise back office systems that are needed to deliver the AMI-specific business functions.

AMI Lighthouse Council
A group of utilities and vendors working with SAP AG to implement AMI programs that are integrated into SAP's enterprise solutions.

AMI-Network Task Force
A subgroup within the UtilityAMI working group of the Open Smart Grid Subcommittee in the UCAIug.

AMI-SEC TF (AMI Security Task Force)
A task force within UCAIug and the Open SG subcommittee charged with developing collaborative security guidelines, recommendations, and best practices for AMI systems. It is in the process of producing technical specifications that may be used by utilities to assess and procure security-related functionality. In addition to utility use, this specification will be used by the OpenAMI task force as part of the AMI/DR Reference Design specification and by vendors to produce compliant and compatible security technologies.

Ampacity
The maximum amperes a conductor such as a line could continuously carry without overheating.

Ampere
The unit of measurement of electrical current produced in a circuit by one volt acting through a resistance of one ohm.

AMR (Automated Meter Reading)
Electricity meters that collect data for billing purposes only and transmit this data one-way from the meter to a meter-reading device. Transmission occurs from radio, PLC, and wireless technologies and is read by the utility resources as they walk or drive by the meters.

AMS (Advanced Meter System)
A collection of smart meters that includes the intelligence to gather meter data at regular intervals and communicate this information to consumers for demand response programs and share this information with internal utility systems like work orders, billing, and outage management.

AMWA (Association of Metropolitan Water Agencies)
An organization of the largest publicly owned drinking water systems in the United States that focuses on federal water policies regarding safe and cost-effective drinking water laws and regulations. (http://www.amwa.net/cs/home)

Anammox® (ANaerobic AMMonium Oxidation)
A wastewater treatment process that uses anammox bacteria to remove nitrogen from wastewater. It replaces nitrification/denitrification processes, thus requiring less energy, producing less sludge, and reducing CO_2 emissions.

Ancillary services
Services that ensure grid reliability and support the transmission of electricity from generation sites to customer loads, managed by a Balancing Authority or ISO/RTO. Ancillary services assist in the integration of utility-scale intermittent renewables into the grid. Ancillary services include spinning reserves, supplemental reserves, and regulation and frequency response. Spinning and non-spinning reserves and regulation are classified as dispatchable DR (demand response). Batteries, flywheels and other advanced energy storage technologies are well suited to provide ancillary services. Traditionally used for transmission but now being considered for use in distribution grids.

Ancillary services market program
A type of demand response program for large electricity customers located within ISOs (Independent System Operators) or RTOs (Regional Transmission Organizations) that conduct markets to accept customer bids for load reductions. If their bids are accepted, customers are paid the market price for committing to be on standby. If their load reductions are needed, they are called by the ISO/RTO and may be paid the spot market energy price.

ANL (Argonne National Laboratory)
DOE (Department of Energy) national laboratory operated by a limited liability corporation associated with the University of Chicago that conducts energy research in several disciplines. The Transportation Technology R&D Center is focused on V2G (Vehicle to Grid) applications and a standard connector for PHEVs. It also conducts research in alternative fuels, fuel cells, and batteries for transportation applications. (http://www.transportation.anl.gov/smart_grid/)

Anode
A negative terminal on a battery.

Anonymize
A term that describes the process of removing personal information from a data set so that it can no longer be associated with a specific individual. It is accomplished using data processing actions to strip data that can identify the individual ownership or originator of specific data. The ability to "de-identify" data is valuable to a variety of entities (utilities, third party service providers, data brokers, governmental agencies) to gain knowledge of aggregated consumer or business behaviors, trends, and preferences that can influence development and delivery of future products, services and policies.

ANSI (American National Standards Institute)
A private, nonprofit organization that oversees the development of voluntary consensus standards for products, services, processes, systems, and personnel in the United States. ANSI accredits NERC's (North American Electric Reliability Corporation) reliability standards development process. The organization also coordinates with the IEC (International Electrotechnical Commission) and is actively engaged in accreditation programs, such as ISO (International Organization for Standardization) 9000 and ISO 14000 for environmental management. (www.ansi.org)

ANSI C12.1
This approved ANSI standard outlines performance criteria and safety tests for electric meters.

ANSI C12.18
This approved ANSI standard for Protocol Specification for ANSI Type 2 Optical Port describes criteria for two-way communication with an electric meter via an optical port. It is sometimes referred to as part of the protocol suite of ANSI standards, including C12.18, C12.19, C12.21, and C12.22.

ANSI C12.19
This approved ANSI standard for Utility Industry End Device Tables defines a table structure for utility application data to be passed between an end

device, such as a meter, and any other device. This standard forms the basis for common industry data structures and provides a common industry language for meter data communications and AMI infrastructures. It is sometimes referred to as part of the protocol suite of ANSI standards, including C12.18, C12.19, C12.21, and C12.22.

ANSI C12.20
This approved ANSI standard for Electricity Meters - 0.2 and 0.5 Accuracy Classes sets performance criteria for meters including accuracy class designations, voltage and frequency ratings, current class designations, test current values, and now allows three phase current and voltage sources as a test method in addition to existing single phase methods.

ANSI C12.21
This approved ANSI standard for Protocol Specification of Telephone Modem Communication is designed for meter communications using a modem and is typically used for AMR (Automated Meter Reading) applications. It is an extension and adaptation of ANSI C12.18 for modems. It is sometimes referred to as part of the protocol suite of ANSI standards, including C12.18, C12.19, C12.21, and C12.22.

ANSI C12.22
This approved ANSI standard for Protocol Specification for Interfacing to Data Communications Networks extends the concepts of ANSI C12.18, ANSI C12.19, and C12.21 standards to allow transport of table data over any reliable networking communications system, including wired or wireless networks. This is relevant to AMI infrastructure. It is sometimes referred to as part of the protocol suite of ANSI standards, including C12.18, C12.19, C12.21, and C12.22.

ANSI C12.23
This standard on AMR Device Compliance Test Standards outlines a series of procedures that validate devices that communicate via the ANSI C12.18 set of standards.

ANSI C12.24
This is a technical report that establishes names and mathematical definitions to calculate Volt-Ampere, Volt-Ampere hours, Volt- Amperes Reactive, and Volt-Ampere Reactive hours formulas in AC electricity meters.

ANSI C2-2007
This ANSI and IEEE standard is also known as the National Electric Safety Code. It addresses human safety during the installation, operation, and maintenance of electric supply and communications lines.

ANSI C57.12.25
This ANSI standard covers the performance, interchangeability, and safety of Pad-Mounted Transformers.

ANSI C57.12.28
This ANSI standard covers the conformance tests and requirements for the integrity of pad-mounted enclosures that contain equipment energized in excess of 600 volts.

ANSI Z535
A collection of 6 standards that define the visual displays or signs to ensure human safety around electrical products. It is also a NEMA (National Electrical Manufacturers Association) standard.

ANSI T1.105
A standard defining synchronous data transmission on optical media. Also known as SONET (Synchronous Optical Network).

ANSI/ASHRAE/USGBC/IES 189.1- 2011

The ANSI/ASHRAE/USGBC/IES Standard 189.1-2011 provides a design standard for high-performance buildings and covers site sustainability, water-use efficiency, energy efficiency, indoor environmental quality, and the building's impact on the atmosphere, materials and resources. The proposal suggests changes for water conservation, including faucet capacities, the use of municipal reclaimed water for irrigation and limiting toilet flushing volumes.

ANSI/CEA 709.1 –b– 2002

This is the standard that defines the communication protocol for networked control systems. The protocol directs peer-to-peer communication for networked protocol, and can also implement master-slave set-ups. The interface in this specification supports multiple types of transmission media. It is part of the LonWorks suite of standards.

ANSI/CEA 709.2-A R-2006

This is the standard that specifies the CEA 709 control network power line (PL) channel. Its objective is to present the necessary information for the development of a physical PL network, as well as nodes to communicate information over the network. This standard also dictates a set of guidelines regarding physical and electrical specifications for the PL system.

ANSI/CEA 709.3 R-2004

This is the standard that specifies the EIA-709.3 free-topology twisted pair channel, and supplements the 709.1 Control Network Protocol specifications. The standard defines communication at 78.125 kbps between multiple nodes along a PL system.

ANSI/CEA 709.4: 1999

This standard defines a seven-layer protocol stack for communications through an EIA-709.4 single fiber, half-duplex fiber optic channel. This

channel enables communications at 1250 kbps between multiple nodes along a PL.

ANT™
A proprietary wireless PAN (Personal Area Network) sensor technology that uses the unlicensed 2.4GHz spectrum for applications that include smart homes and industrial uses.

APEC (Asia Pacific Economic Cooperation)
An inter-governmental organization that operates on consensus and non-binding decisions that released a data privacy framework and cross-border data privacy rules for its members to consider to provide privacy protection while avoiding barriers to information flows. (http://www.apec.org/)

APPA (American Public Power Association)
A nonprofit organization for more than 2,000 not-for-profit, state or community-owned, or municipal electric utilities that serve over 45 million Americans. It provides advocacy, education, and other services to members to ensure reliable electricity service at competitive costs. Its subsidiary, Hometown Connections, secures group discounts for AMR (Automated Meter Reading), SCADA (Supervisory Control And Data Acquisition), BPL (Broadband over Powerline), and CIS (Customer Information System) solutions. (www.appanet.org)

Apparent energy
The integral of apparent power with respect to time, usually measured in kVAh (kiloVolt-Ampere-hours)

Apparent power
The product of the voltage and the current (amperes). It contains both active and reactive power. It is measured in volt-amperes and often expressed in kVA (kilovolt-amperes) or MVA (megavolt-amperes).

Appliance efficiency index
A comparative measure of trends in appliance efficiencies for new major appliances and energy-using equipment. The base year for relative comparisons was 1972 (1972=100).

Appliance efficiency standards
A set of minimum efficiency standards for major home appliances, including furnaces, air conditioners, refrigerators, freezers, water heaters, dishwashers, and heat pumps, established by the National Appliance Energy Conservation Act of 1987.

APU (Ancillary Power Unit)
Units designed to provide grid stabilization and backup services by storing energy until it is needed by the grid in the event of a power plant or other asset failure.

APWA (American Public Works Association)
A nonprofit international educational and professional association of public agencies, private sector companies, and individuals dedicated to providing high quality public works goods and services. (http://www.apwa.net/)

Aqueduct
A constructed conveyance for water.

Aquifer
A groundwater or underground source of raw water that may or may not require pumping to be accessible for consumption.

AREM (Alliance for Retail Energy Markets)
A not-for-profit corporation of electric service providers, active in California's retail electricity market, promoting direct access and advocacy in state policy. Members include Commercial Energy, Constellation

NewEnergy, Direct Energy, Reliant Energy, and Sempra Energy Solutions. (http://www.retailenergymarkets.com/)

ARIB (Association of Radio Industries and Businesses)
A Japanese-based industry organization that promotes radio technologies, radio spectrum use, and standards. (http://www.arib.or.jp/english/html/arib/index.html)

ARPA-E (Advanced Research Projects Agency – Energy)
An agency within the DOE (Department of Energy) with the mission to fund projects that will develop transformational technologies that reduce America's dependence on foreign energy imports; reduce USA energy-related emissions (including greenhouse gases); improve energy efficiency across all sectors of the USA economy and ensure that the USA maintains its leadership in developing and deploying advanced energy technologies. (http://arpa-e.energy.gov/)

ARPU (Average Revenue Per User)
A standard metric used by service providers that is calculated by dividing revenue by the total number of users who provide that revenue. It is also known as average revenue per unit and is useful to gauge success of subscription-based companies.

ARRA (American Recovery and Reinvestment Act)
Official name for the 2009 Stimulus Bill enacted by the federal government in 2009. It contains about $40 billion in funding and loan guarantees related to Smart Grid solutions, including R&D for energy efficiency and traction batteries for electric vehicles. There is a section of the Bill, "Electricity Delivery and Energy Reliability," that is referred to as the Smart Grid provision. It also provided funding for the Clean Water State Revolving Fund (SRF) and the Drinking Water SRF and for water recycling projects under the guidance of the United States Bureau of Reclamation (part of the US Department of the Interior.)

ARRA Demonstration Grant
A DOE (Department of Energy) investment program that supports demonstrations of the innovative use of emerging technologies in the power grid, with a maximum award of $100 million. It is aimed at developing new and cost-effective Smart Grid equipment, tools, techniques, and system configurations that significantly improve upon today's technologies.

ARRA Investment Grant
A DOE (Department of Energy) (program intended to enable Smart Grid functions on the nation's electrical grid as soon as possible. The cost-shared grants support manufacturing, purchasing, and installation of existing Smart Grid technologies that can be deployed on a commercial scale, with a maximum award of $200 million. Grants given to corporations under this program were ruled by the IRS to be exempt from federal tax.

Array
PV (Photovoltaic) panels connected together electrically to form a single electrical output.

ASAI (Average System Availability Index)
A commonly used measurement of electric utility reliability. It is determined by the number of customer hours available, divided by the number of customer hours served. Also known as Average Service Availability Index.

ASAP-SG (Advanced Security Acceleration Project – Smart Grid)
A task force comprised of North American utilities, UCAIUG UtiliSec WG, and governmental entities like NIST (National Institute of Standards and Technology) to develop a Smart Grid Security Profile Blueprint. The profiles will offer guidance for building and implement Smart Grid security, and an independent team of security experts will conduct a usability and gap analysis on the blueprint and profiles.

ASCE (American Society of Civil Engineers)
An organization promoting technology, environmental stewardship, and professional education in the civil engineering profession. (http://www.asce.org/)

ASD (Adjustable Speed Drives)
Drives that save energy by ensuring the motor's speed is properly matched to the load placed on the motor. Related terms include polyphase motors, motor oversizing, and motor rewinding.

ASDWA (Association of State Drinking Water Administrators)
An association assisting states in meeting their obligation to provide safe drinking water as required by the SDWA and providing information and expertise to create drinking water programs. (http://www.asdwa.org/)

ASE (Alliance to Save Energy)
A global nonprofit coalition of business, government, environmental, and consumer leaders founded in 1977 that promotes energy efficiency for a healthier economy, a cleaner environment, and greater energy security. It conducts initiatives in research, policy advocacy, and education; board-level activism; and encourages public-private partnerships. Initiatives include Commercial Buildings, Industrial Energy Efficiency, and Energy Efficient Codes. (http://ase.org/)

ASERTTI (Association of State Energy Research and Technology Transfer Institutions)
A nonprofit organization working to increase the effectiveness of energy research efforts in contribution to economic growth, environmental quality, and energy security between state energy RD&D (Research, Development, and Deployment) organizations. Members include state and federal agencies, universities, and private corporations. (www.asertti.org/)

ASHRAE (American Society of Heating, Refrigerating, and Air-Conditioning Engineers)
A nonprofit technical organization whose 51,000 members influence the direction of heating, ventilation, air-conditioning, and refrigeration technology through industry standards and recommended procedures and guidelines, research, and technical information. Areas of expertise include energy efficiency, green building design, and data center AC and ventilation. (http://www.ashrae.org/)

ASHRAE 90.1
A standard for energy efficiency in buildings excluding low-rise residential structures. It covers building envelope, HVAC, and lighting. It has been a benchmark for US commercial building energy codes and influenced international codes and standards.

ASHRAE 135-2008
A standard that defines a data communication protocol for building automation solutions that allows building equipment and systems manufactured by different companies to work together. It includes information about a load control object to enable more integration between utilities and buildings in future demand response programs.

ASHRAE 189.1
A standard that defines the minimum requirements for siting, design, construction, and plan for high performance green building operations. It addresses site sustainability, water use efficiency, energy efficiency, indoor environmental quality, and building impacts on the environment.

ASIWPCA (Association of State & Interstate Water Pollution Control Administrators)
A professional organization consisting of the state, interstate and territorial officials responsible for the implementation of surface water protection programs throughout the USA. Its mission is to protect and restore

watersheds to ensure clean water through best practices and public policy. (http://www.asiwpca.org/)

ASME (American Society of Mechanical Engineers)
A not-for-profit professional organization that promotes the art, science, and practice of mechanical and multidisciplinary engineering and allied sciences throughout the world. (http://www.asme.org/)

ASP (Application Service Provider)
Any third party entity that owns a software application and manages it to be available to their customers to use over a network. Also known as Software as a Service (SaaS).

ASPRS (American Society for Photogrammetry and Remote Sensing)
A global, scientific association with over 7,000 members. Its mission is to advance knowledge of mapping sciences and promote responsible applications of imaging, remote sensing, GIS (Geographic Information System), and supporting technologies. (http://www.asprs.org/)

ASTM International
Formerly known as the American Society for Testing and Materials (ASTM), a not-for-profit entity focused on development and publication of international voluntary consensus standards for materials, products, systems and services. Members include producers, users, consumers, government, and academia from over 135 countries. Its water testing standards allow water managers to test the quality of water for safe consumption. (http://www.astm.org/)

ASV (Automatic Shutoff Valve)
A valve with actuators to open or close the valve based on data sent to the actuator from natural gas pipeline sensors. Human intervention is not required to operate the valve.

ATIS (Alliance for Telecommunications Industry Solutions)
An industry organization that develops standards for the ICT industry with focus on IP infrastructure, converged multimedia services, and enhanced OSS (Operations Support Systems) and BSS (Business Support Systems). (http://www.atis.org/)

ATM (Asynchronous Transfer Mode)
A high-speed switching technology typically found in digital backbone networks. It organizes data into 53-byte packets that are reassembled at the receiving node, which has a dedicated connection for the duration of transmission. Network speeds can reach 10 Gbps.

Attack surface
Security term to describe the vulnerability of an application or system. Attack surface increases with remote management capabilities.

Attenuation
The loss of signal strength in a radio signal or electrical signal that is typically related to the distance the signal travels.

Authentication
A process to verify an identity. It can be as simple as a password or more complex biometrics such as fingerprint or retina scans.

Automated capacitor controls
Remote control of capacitor banks to switch them on and off as needed, instead of manually controlled on predictive time schedules. It is part of distribution automation.

Automated DR (Automated Demand Response)
Automated systems typically used by C&I (Commercial and Industrial) customers to reduce energy use to save money, reduce load during peak

periods, and get credits for emission reductions. It is also known as Intelligent DR.

Automated recloser
Any device that opens when it senses a fault, and automatically closes according to a defined time and/or condition to restore service. Commonly found in distribution networks, these devices help utilities maintain reliable electricity delivery and lower costs by reducing manual adjustments and repairs.

AUTOSAR (AUTomotive Open System ARchitecture)
An industry consortium of car manufacturers, suppliers and other companies from the electronics, semiconductor and software business sectors. Their focus is on standardization of basic system functions and functional interfaces, the ability to integrate, exchange and transfer functions within a car network and to substantially improve software updates and upgrades over the vehicle lifetime. (http://www.autosar.org/)

Auxiliary generator
Any electric plant generator that powers the electrical generating equipment or plant lighting and other building systems during times when the plant is offline or power is unavailable from the grid.

Availability
A term to describe reliability and is calculated as the number of hours a resource is available to provide power divided by the total hours in the year.

AVL (Automatic Vehicle Location)
The capability to determine a vehicle's geographic location using GPS or terrestrial radio positioning systems and used to track utility field crews.

Avoided capacity
The amount of capacity that can be displaced but still meet system reliability requirements. DR (Demand Response) programs contribute reductions in energy use that can be calculated as avoided capacity.

Avoided cost
A price calculation based on the amount of money that a utility avoids paying for generation through use of cogeneration or distributed generation facilities. It is the marginal cost for a regulated utility to generate or purchase one more unit of power.

AVVC (Advanced Volt-VAR Control)
Realtime applications that automatically coordinate voltage control and VAR losses across a distribution grid using advanced algorithms to model optimal load flows. Also known as VVO (Volt-VAR Optimization).

AWE (Alliance for Water Efficiency)
A nonprofit organization dedicated to the efficient and sustainable use of water and advocates for water efficiency, education and research. Members include equipment manufacturers, and environmental, academic and government groups. (http://www.allianceforwaterefficiency.org/)

AWEA (American Wind Energy Association)
A national trade association of America's wind industry consisting of over 2,500 member companies covering wind generation, manufacturing, and service suppliers. Its mission is to promote wind power growth through advocacy, communication, and education. (http://www.awea.org/)

AWRA (American Water Resources Association)
A nonprofit professional association dedicated to the advancement of water resources management, education, and research. Members include engineers, foresters, biologists, educators, ecologists, geographers, regulators, hydrologists and attorneys. (http://www.awra.org/)

AWT (Association of Water Technologies)
An international nonprofit trade organization representing about 400 companies specializing in water treatments for industrial and commercial cooling and heating systems. It delivers business and professional education and resources to its members, including training, certification, and regulatory and public awareness programs. (http://www.awt.org/)

AWWA (American Water Works Association)
An international nonprofit organization of over 4,600 water utilities that provide knowledge, information and advocacy to improve the quality and supply of water in North America. (http://www.awwa.org)

B

B&P (Business and Policy)
A working group within NIST (National Institute of Standards and Technology) focused on connecting its technical work to higher-level business and political issues. (http://www.nist.gov/smartgrid/)

B2G (Building to Grid)
A shorthand term for the focus on commercial building integration to the grid. It is also the name for a specific working group within NIST (National Institute of Standards and Technology) focused on this subject. (www.nist.gov/smartgrid/)

Backfeed
The ability to flow electricity back to a utility.

Backflow
A hydraulic condition that causes water to flow in the opposite (and unwanted) direction in unprotected drinking water systems.

Backhaul Network
Wired or wireless, the links that carry traffic from nodes or sites back to the Internet or other main network backbone. In AMI (Advanced Metering Infrastructure), the backhaul network could be the communication facilities that link substations back to utility control centers.

Backup Generator
A generator that is used only for test purposes or in an emergency, such as a shortage of power needed to meet customer load requirements.

Backup power
Electric energy supplied by a utility to replace power and energy lost during an unscheduled equipment outage.

Backwash
A process to clean membrane filters used for treating potable water. It can use liquid or gas to clean large to small-scale water utility facilities.

BACnet® (Building Automation and Control net)
A data communication protocol for building automation-and-control networks. Also known as SSPC 135. Developed under the auspices of the American Society of Heating, Refrigerating and Air-Conditioning Engineers (ASHRAE), BACnet is also an ISO standard – ISO 16484-5. NIST's (National Institute of Standards and Technology) Mechanical Systems and Control Group is working with the utility industry to develop technical standards for automated communications to enable the building of automation system participation in the Smart Grid. NIST has also been working with BACnet teams to develop this protocol to support distributed load control and meter reading.

BACnet® International (Building Automation and Control net International)
An industry organization that promotes BACnet in building automation and control systems through interoperability testing, educational programs, and promotional activities. Members include companies that design, manufacture, install, commission, and maintain equipment that uses BACnet for communication. Their test facilities conduct compliance and interoperability testing and award the BACnet Testing Laboratory (BTL) symbol to products that conform to the BACnet standard. (http://www.bacnetinternational.org/)

BACnet® NS-WG (BACnet Network Security Working Group)
A team focused on addressing secure network configurations that can be implemented now and on identifying changes to the BACnet standard that will allow for secure message exchange.
(www.bfrl.nist.gov/863/BACnetNSoverview.html)

BACnet® TL-WG (BACnet Testing Laboratory Working Group)
A volunteer team that provides oversight and guidance to the Testing Laboratories and conformance certification programs.

BACnet® UI-WG (BACnet Utility Integration Working Group)
A NIST (National Institute of Standards and Technology) team working with Lawrence Berkeley National Lab's DRRC (Demand Response Research Center) and other partners on a standard for automated demand response to allow buildings to receive and react to Web-based messages from utilities. Messages include critical event and other notifications, price signals, automated bidding for generation (or for load shedding), meter reading, and energy-usage history. This will allow energy service providers to interact with any building regardless of which vendor's building automation system is in use, and likewise let any building communicate with any utility without a utility-specific interface. The basis for the communications is BACnet Web Services.
(www.bfrl.nist.gov/863/BACnetUIoverview.html)

Bakersfield effect
A negative consumer reaction to any Smart Grid technology rollout. It may trigger project delays and/or additional investments in consumer outreach and education.

Balancing area
A geographic area, formerly known as a control area, in which the balancing authority matches generation with customer demand, and the transmission operator monitors the flows over the transmission system and voltages at substations. It is one of the regional functions contributing to the reliable planning and operation of the bulk power system. These areas are connected to each other by tie lines.

Balancing authority
The authority that integrates resource plans ahead of time and maintains real-time balancing of electricity resources and electricity demand for a region with common generation-control operations.

Ball valve
A type of valve generally used for throttling clear water.

BANANA (Build Absolutely Nothing Anywhere Near Anything)
An upgrade to NIMBY (Not In My Backyard), with ramifications about placement of new transmission lines, generating facilities (whether fossil fuel or renewables), and other facilities that are part of a local, regional, or national grid.

Bandwidth
The information capacity of a communications resource, usually measured in bits per second.

Barometer
A type of sensor that detects air pressure

BAS (Building Automation Systems)
Software and hardware deployed in buildings to automatically manage refrigeration, HVAC, lighting and other building energy usage on a continuous basis. The objective of BAS is to deliver optimal occupant comfort while minimizing energy use. These control systems are the integrating components to fans, pumps, heating/cooling equipment, dampers, mixing boxes, and thermostats. Monitoring and optimizing temperature, pressure, humidity, and flow rates are key functions of modern building control systems. Also known as EMCS (Energy Management and Control Systems) or Energy Management Systems (EMS).

Baseload
The minimum amount of electric power required to meet demand for a given period of time at a steady rate, or the portion of the electricity demand that is continuous and does not vary over a 24-hour period.

Baseload capacity
The generating equipment normally operated to serve loads on an around-the-clock basis.

Baseload plant
A generating plant in continuous operation that contributes to system load, producing electricity at an essentially constant rate. These plants are run to maximize system mechanical and thermal efficiency and minimize system operating costs.

Base rate
A fixed kilowatt hour charge for electricity consumed.

Base station
A land station in a mobile service, which interacts with mobile devices, other base stations, and the public switched telecommunications network.

Battelle
A nonprofit organization that develops and commercializes technology and manages laboratories, including Pacific Northwest National Laboratory, Brookhaven National Laboratory, Oak Ridge National Laboratory, and NREL (National Renewable Energy Laboratory). Key energy initiatives include developing commercially viable fuel cells and energy storage. (http://www.battelle.org/)

BAU (Business-As-Usual)
A term used in models and projection exercises to illustrate consumer behaviors around energy consumption without any introduced changes in technology, policy, or procedure. It describes the status quo.

BCXA (Building Commissioning Association)
An international nonprofit organization focused on commercial buildings industry commissioning processes. It conducts education and manages a certification program for credentialed commissioning professionals. (http://www.bcxa.org/)

BEA (Battelle Energy Alliance)
A Battelle-led industry team that is managing INL (Idaho National Laboratory). The team consists of BWX Technologies, Inc., Washington Group International, EPRI (Electric Power Research Institute), and an alliance of university collaborators, including Massachusetts Institute of Technology (MIT) and several nuclear engineering universities.

BEEP (BOMA Energy Efficiency Program)
A training program created in partnership with the EPA ENERGY STAR program. It teaches property owners, managers, and operators how to reduce energy consumption and costs with proven strategies for optimizing equipment, people, and practices. The commercial office building industry spends approximately $24 billion annually on energy. A 30-percent reduction in energy consumption ($7.2 billion savings) is achievable simply by improving building operating standards. (http://www.boma.org/TrainingAndEducation/BEEP/Pages/default.aspx)

BES (Bulk Electric System)
The part of the overall electricity system that includes the generation of electricity and the transmission of electricity over high-voltage transmission lines to distribution companies. This includes power-generation facilities, transmission lines, interconnections between neighboring transmission

systems, and associated equipment. It does not include the local distribution of the electricity to homes and businesses. Also known as BPS (Bulk Power System).

BESS (Battery Energy Storage System)
An integrated system including a battery pack, battery management system, power conversion system and appropriate balance of plant required to operate the system.

BEV (Battery Electric Vehicle)
An electric vehicle that uses chemical energy that is stored in rechargeable battery packs.

Beyond the meter
An expression that acknowledges the traditional electrical grid end point (meter) at a building, and assumes use of new Smart Grid-enabled technologies to impact electricity consumption inside home or commercial building premises.

Bi-directional communication
Two-way communication conducted by wired or wireless media. It is an important feature of the Smart Grid, and can use telecom networks and/or electrical grids to convey information.

Bi-directional electricity
Two-way electricity flow managed in distribution and transmission grids. It is an important feature of the Smart Grid, and enables integration of distributed energy resources into local and regional grids for flexibility and improved reliability.

Billing period
The time period between meter readings.

Binary plant
A type of geothermal generating technology that uses two fluids to create electricity – the hot water from the geothermal source, and another liquid like isobutene with a lower boiling point that vaporizes to steam to drive a generator. It is typically used in geothermal sources with temperatures lower than 150 degrees Celsius or 302 degrees Fahrenheit. The other types of geothermal plants are flash power plant, dry steam power plant, and flash/binary combined cycle.

Biomass energy
As defined in the ARRA Section 1306 amending EISA 2007 (Energy Independence and Security Act of 2007), renewable energy derived from plant materials or animal waste. Examples include food crops, grassy and woody plants, algae, and landfill gas (methane).

Bioterrorism Preparedness and Response Act of 2002
Federal legislation that requires every water system that serves more than 3,300 people to perform a vulnerability assessment and report results to the EPA. These systems must develop an emergency response plan to address identified vulnerabilities within 6 months of completion of the assessment.

BIPV (Building Integrated Photovoltaic)
Solar technologies that are part of the building structure, most often the roofing system. Also known as BIP.

Black water
Wastewater that contains food, animal, or human waste coming from toilets, kitchen sinks, and dishwashers.

Blackstart capability plan
A documented procedure for a generating unit or station to go from a shut-down condition to an operating condition delivering electric power without

assistance from the electric system. This procedure is only a portion of an overall system restoration plan.

BLE (Bluetooth Low Energy)
Also known as Bluetooth Smart, this wireless personal area network technology uses the same communications protocols as Bluetooth, but at reduced power consumption and cost, with enhanced range. These properties enable an expansion into the IoT space.

Block-rate structure
A rate schedule that charges a different cost for increasing blocks of demand for energy.

Blowdown
Water drained from cooling towers and steam or hot water boiler systems to prevent mineral buildups. Also known as bleed water.

Bluetooth®
A PAN (Personal Area Network) wireless technology that works up to a distance of 50 meters or 164 feet. It operates in the unlicensed 2.4 – 2.485 GHz spectrum.

Bluetooth® Special Interest Group
A not-for-profit industry association that publishes technical specifications, manages the qualification program, and promotes Bluetooth technology. (https://www.bluetooth.org/apps/content/)

BNL (Brookhaven National Laboratory)
A DOE (Department of Energy) laboratory that will conduct energy-related R&D in the effective uses of renewable energy through improved conversion, transmission, and storage. The Energy Sciences and Technology Department and the Superconducting Magnet Division are two groups with research objectives targeted to Smart Grid applications. BNL is also working

on DOE's Water-Energy Nexus Initiative, applying a multi-disciplinary approach to solving interconnected water and energy problems. (http://www.bnl.gov/world/)

BOMA (Building Owner and Managers Association)
An industry association that manages more than 9 billion square feet of commercial properties in North America. It provides information on office building development and operating costs, energy consumption, local and national building codes, legislation, and technological developments. Members are building owners, managers, developers, corporate facility managers, and industry vendors. It runs the BEEP training program and recognizes environmentally friendly buildings with its Annual Earth Award. (http://www.boma.org/Pages/default.aspx)

Booster station
A facility that ensures that water retains the required level of purity through transport from treatment plant to end user.

BOT (Build Own Transfer)
A type of financing for projects, where a private entity receives a construction contract from the private or public sectors to finance, design, construct and finally operate the facility specified in the contract before transferring the assets back to the investing sector. This strategy is used for the construction of new telecommunications infrastructure, water projects, and energy generating plants (including renewables).

BPA (Bonneville Power Administration)
A federal entity in DOE (Department of Energy) that markets wholesale electricity and transmission to the Pacific Northwest's public and private utilities and some large industries. BPA provides about half the electricity used in the Northwest and operates over three-fourths of the region's high-voltage transmission. (http://www.bpa.gov/corporate/)

BPI (Building Performance Institute) A not-for-profit national standards development and credentialing organization for residential energy efficiency retrofit and weatherization work. (http://www.bpi.org/)

BPL (Broadband over Power Line)
A communications technology that uses the electric power line to send and receive high-speed data to devices plugged in to receive electricity. BPL can signal devices to cycle on and off or shut down as needed.

BPS (Bulk Power System)
The part of the overall electricity system that includes the generation of electricity and the transmission of electricity over high-voltage transmission lines to distribution companies. This includes power-generation facilities, transmission lines, interconnections between neighboring transmission systems, and associated equipment. It does not include the local distribution of the electricity to homes and businesses. Also known as BES.

Brackish water
Water with a salinity between seawater and freshwater. It requires treatment for human consumption.

Brayton cycle
A thermoelectric generation process in which fuel and a compressor heat and increase a gas pressure that expands over a turbine. As the turbine blades spin, the generator, a coiled wire cylinder that spins in a magnetic field, generates electricity. It is used with "clean" fuels such as natural gas, synthetic gas from coal, or biomass gasification.

Breakthrough technology
As defined in the ARRA Section 1306 amending EISA 2007 (Energy Independence and Security Act of 2007), any major advancement in an engineering field that accelerates the effectiveness, adoption, and results of solutions.

Brine stream
Wastewater resulting from a desalination process. It is very salty and full of minerals and organic contaminants.

Broadband
Information capacity or bandwidth that is higher than 2 Mbps to manage multiple voice, data, and video channels at the same time.

Broadband Forum
An industry organization that develops multi-service broadband packet networking specifications for interoperability, architecture and management of broadband networks that cover in-home to backbone uses. Their specifications are TRs (Technical Reports) and there are quite a number that apply to the consumer edge of the Smart Grid. (http://www.broadband-forum.org/)

Brownout
An intentional voltage reduction on the grid that reduces load during shortages that can be caused by peak demand timeframes or loss of generation resources.

BRS (Broadband Radio Service)
A commercial broadband wireless point-to-multipoint specification that uses UHF (Ultra High Frequency) communications. It operates on FCC (Federal Communications Commission)-licensed frequencies. It was formerly known as MMDS (Multi-channel Multi-point Distribution System).

BSC (Base Station Controller)
A device that controls a number of base station transceivers in a wireless network.

BSS (Base Station System)
A fixed wireless network device that enables communications with other fixed or mobile devices. From a geographic applications perspective, it also "fixes" latitude and longitude for satellite-based global positioning systems (GPS).

BTI (Breakthrough Technologies Institute)
A nonprofit educational organization that identifies and promotes environmental and energy technologies. It was established in 1993 to promote emerging technologies in environmental and energy policy debates. Current education programs focus on fuel cell education, innovative mass transit strategies, and air-quality planning and pollution control. FC2K (Fuel Cell 2000) is one of its main activities. (http://www.btionline.org/)

BTL (BACnet® Testing Laboratory)
Test facilities and procedures run by BACnet International to ensure vendor compliance and interoperability to the BACnet standards. Tested products receive a BTL mark that indicates the products correctly implement the BACnet features claimed in the listing.
(http://www.bacnetinternational.net/btl/)

BTS (Base Station Transceiver)
A network device that communicates with other devices in the wireless network and controllers.

BTU (British Thermal Unit)
The quantity of heat required to raise the temperature of 1 pound of liquid water by 1 degree Fahrenheit. It is also defined as the temperature at which water has its greatest density, which is approximately 39 degrees Fahrenheit.

Budget plan
An agreement between a customer and a utility that allows the household to pay the same amount for energy for each month for a number of months.

BUGS (Backup Generation Sources)
Generators that are required by building codes to operate elevators, fire systems, and critical services during emergencies. Backup generation sources are usually separately activated in an emergency, but, if retrofitted with remote sensing and management technology, could play a role as distributed generation sources.

Building shell conservation feature
A building feature designed to reduce energy loss or gain through the shell or envelope of the building. Features include roof, ceiling, or wall insulation; insulated windows; tinted or reflective glass or shading films; exterior or interior shadings or awnings; and weather stripping or caulking.

Bulk power transactions
Wholesale transactions between electric utilities for situations ranging from maintaining load to cost reductions.

Bulk water supplier
An entity that sells treated or untreated water to a variety of users, including utilities. Generally located in the southwest USA, these suppliers also manage large aqueducts.

Bundled utility service
Energy generation, transmission, distribution, and ancillary functions are provided by one utility.

Bus
An electrical conductor that is a common connection between two or more electrical circuits.

Butterfly valve
A valve generally used for handling large flows of gases or liquids, including slurries.

BWR (Boiling Water Reactor)
A light-water reactor in which water, used as both coolant and moderator, is allowed to boil in the core. The resulting steam can be used directly to drive a turbine and generate electricity.

C

C&I (Commercial and Industrial)
A common customer designation to distinguish these customer segments from residential energy customers. C&I customers typically have different meters, different rates, and different DR programs to reduce energy use during peak demand periods.

CABA (Continental Automated Building Association)
A not-for-profit international industry association headquartered in Canada, focused on intelligent home and building technologies in North America. Members include utilities, government agencies, and those involved in the design, manufacture, installation, and sales of home and building automation products. It encourages the development of industry standards and protocols. (http://www.caba.org/)

CAC (Cold Aisle Containment)
A data center practice of physically separating hot exhaust air and cold air by covering an aisle so that cold air passes through perforated floor tiles to server rack air inlets.

CAES (Compressed Air Energy Storage)
An energy storage solution based on the compression of air to be used later as an energy source. Utilities can store compressed air during periods of low energy demand for use in peak load conditions. Compressed air is a bulk storage solution that may also be used with renewable energies.

CAIDI (Customer Average Interruption Duration Index)
An IEEE (Institute of Electrical and Electronics Engineers) reliability index to track and benchmark reliability performance in utilities. It is calculated by taking the SAIDI (System Average Interruption Duration Index) measurement and dividing it by the SAIFI (System Average Interruption Frequency Index) measurement.

CAISO (California Independent System Operator)
A not-for-profit corporation established in 1998 charged with operating the majority of California's high-voltage wholesale power grid. This ISO (Independent System Operator) is the link between power plants and the utilities, provides equal access to the grid for all qualified users, and promotes infrastructure development. Its mission statement includes environmental stewardship along with reliable grid operations and market effectiveness. Its latest strategic plan identifies integration of renewable energy sources and Smart Grid technologies into the grid to improve reliability and conform to California energy and air and water quality mandates. The territory covers California and northern Baja California in Mexico. It is one of the 10 ISO/RTOs (Regional Transmission Organization) currently operating in North America. Also known as Cal ISO. (www.caiso.com)

CalCEF (The California Clean Energy Fund)
A nonprofit corporation formed in 2004 to invest in California's clean energy ecosystem. CalCEF programs provide financial and intellectual capital to promising clean energy companies at various investment life cycle stages. (http://www.calcef.org/about.htm)

CalCEF Angel Fund
A for-profit limited partnership. CalCEF is the founding limited partner of the Angel Fund. It invests in seed or start-up businesses and will consider investments throughout the United States, not just California. (http://www.calcefangelfund.com/)

CalConnect^{SM}
A not-for-profit consortium of vendors of calendaring and scheduling systems and tools. The mission of CalConnect is to improve the interoperability of calendaring and scheduling systems through the development of standards, interoperability testing methods, and industry collaboration.

CALMAC (California Measurement Advisory Council)
An organization that provides an unofficial forum for development, implementation, presentation, discussion, and review of market assessment and evaluation studies for energy programs within California. Members are the four California investor-owned utilities (Pacific Gas and Electric, Southern California Edison, Southern California Gas, and San Diego Gas and Electric) and the two state energy agencies -the CEC (California Energy Commission) and the CPUC (California Public Utilities Commission) Energy Division. It has a sister organization that handles DSM (Demand Side Management) programs fielded up to 1997 known as CADMAC. (http://www.calmac.org/)

CAMPUT (Canadian Association of Members of Public Utility Tribunals)
A nonprofit organization of federal, provincial, and territorial boards and commissions, responsible for the regulation of the electric, water, gas, and pipeline utilities in Canada. Its purpose is to improve public utility regulation and provide education and communication opportunities for its members. FERC (Federal Energy Regulatory Commission) in the USA is an associate member of CAMPUT. (www.camput.org/public_index.html)

CAN (Car Area Network)
A network consisting of all electric and electronic control units, sensors, or other components that are interconnected in a car. Also known as a Controller Area Network.

CANENA^{TM} (Council for Harmonization of Electrotechnical Standards of the Nations of the Americas)
A hemispheric industry association focused on electrotechnical standards harmonization activities within the Americas. Activities also include conformity assessment, compliance issues, compatibility, interchangeability, interoperability, installation codes, energy efficiency, and intellectual property topics. (http://www.canena.org/)

CANGEA (Canadian Geothermal Energy Association)
An industry association promoting exploration and development of geothermal resources to supply energy to Canadians and to USA export markets. (http://www.cangea.ca/)

CANWEA (Canadian Wind Energy Association)
A nonprofit trade association that promotes development and application of wind energy in Canada. Members include wind energy owners, operators, manufacturers, project developers, consultants, and service providers. (http://www.canwea.ca)

Cap and Trade
A market based carbon emission limiting policy tool. It works by setting a total "cap" or hard limit on carbon emissions. Once this limit is in place, permits to emit carbon are sold to carbon sources, with the number of permits issued not exceeding the cap. With this system in place, a company that reduces its emissions can profit by selling its unneeded permits to a more heavily polluting company, while reducing net carbon released into the atmosphere.

Capability
The maximum load that a generating unit, generating station, or other electrical apparatus can carry under specified conditions for a given period of time without exceeding approved limits of temperature and stress.

Capacitor
A two terminal electrical component that is used to store an electric charge, consisting of a pair of conductors separated by an insulator. It helps to manage reactive losses and support voltage consistency.

Capacity
The rated continuous load-carrying ability, expressed in MW or MVA of generation, transmission, or other electrical equipment. It is also the

amount of electric power delivered or required for which a generator, turbine, transformer, transmission circuit, station, or system is rated by the manufacturer. Some capacity is also considered dispatchable DR (Demand Response). In a battery context for EV (Electric Vehicles) and PHEVs (Plugin Hybrid Electric Vehicles), it is the total ampere hours drawn from a fully charged cell or battery in specified operating conditions that include discharge rate and temperature.

Capacity charge
A two-part pricing method used in capacity transactions, in which energy charge is the other element. The capacity charge is based on the amount of capacity being purchased. It is sometimes called demand charge.

Capacity factor
The ratio of the electrical energy produced by a generating unit for the period of time considered to the electrical energy that could have been produced at continuous full power operation during the same period. It is calculated as the KWh of electricity generated divided by the total of the KW of generating capacity multiplied by a time period expressed in hours.

Capacity market programs
Types of demand response programs that use customer load reductions as system capacity to replace conventional generation or delivery resources. Customers are notified of needs to reduce loads and incur penalties for any lack of curtailment. Incentives include guaranteed payments for curtailment. These programs occur where ISO/RTOs (Independent System Operators/Regional Transmission Organizations) have ICAP (Installed Capacity) markets and are most often found in wholesale markets.

Capacity transaction
The acquisition of a specified quantity of generating capacity from another utility for a specified period of time.

Capital cost
The one-time set-up costs of field development and plant construction and the equipment required for generation, transmission, and distribution of electricity. These costs are typically recovered through demand charges.

Capital stock
Property, plant, and equipment used in the production, processing, and distribution of electricity.

Car Connectivity Consortium
An industry association focused on developing global standards for smartphone in-car connectivity. They promote technical specifications, test tools for certifying products, and support application developers with user-interface guidelines and conferences. Their first standard is MirrorLink™. (http://www.carconnectivity.org/)

CARB (California Air Resources Board)
As part of the California Environmental Protection Agency, this board reports directly to the California Governor's Office. Its mission is to promote and protect public health and welfare and ecological resources through the effective and efficient reduction of air pollutants while recognizing and considering the effects on the economy of the state. A major goal is to reduce California's green house gas emissions. (http://www.arb.ca.gov/homepage.htm)

Carbitrage
The capability for an EV (Electric Vehicle) or PHEV (Plugin Hybrid Electric Vehicle) or charging station to communicate with the electrical grid to schedule charge/discharge activities based on conditions including pricing signals, tariff agreements, TOU (Time Of Use), DR (Demand Response) programs, and manual overrides by car owners.

Carbon Credit

The equivalent of one metric ton of carbon dioxide that is prevented from entering the atmosphere. It has monetary value based on the type and location of emission reductions. Carbon Dioxide (CO_2) is the baseline credit but six other greenhouse gases (GHGs) can also be credited in reduction strategies.

Carbon Footprint

A measure of human impact on the environment that calculates the amount of greenhouse gases produced in burning fossil fuels for electricity, heating, transportation, agriculture, and all other human activities. It is usually calculated over a year of time, but can reflect different time spans.

Carbon Offset

A financial transaction in which a successful emissions reduction that is produced by one entity can be sold and bought by another entity to reduce its carbon footprint.

Carbon Price

A financial value placed on carbon emissions to address the economic externalities that the market does not currently assign to fossil fuels, including environmental and health degradations.

Carbon Sequestration

The storage of captured carbon emissions. The two primary methods are injection into geologic formations or terrestrial applications.

Carbon Sink

An ecosystem service provided by nature that sequesters or biologically modifies atmospheric carbon. Examples include oceans and forests, which absorb atmospheric carbon, and in the case of aquatic and terrestrial plants, release oxygen.

Carbon Source
Anything that emits CO2 into Earth's atmosphere. This includes man-made sources such as the burning of fossil fuels to natural sources such as forest fires.

CASAGRAS (Coordination And Support Action for Global RFID-related Activities and Standardisation)
A project that is part of the European FP7 initiative that is focused on foundation studies to define and accommodate international issues and developments about RFID and the Internet of Things. (http://www.iot.eu.com/)

CASHEM (Context Aware Smart Home Energy Manager)
A DOE (Department of Energy) project with the goal of creating an integrated home energy management solution to help consumers to manage their home's energy load. It accomplishes these goals via an in-home server, Wi-Fi connected electric devices, and a unified HTML 5 tablet interface. It is a step towards transactive energy.

Casing
Metal or plastic pipe installed in the drilled hole of a well to prevent collapse of the bore hold and entrance of surface pollutants. It also enables placement of pumping equipment.

Cathode
A positive terminal on a battery.

CBECS (Commercial Building Energy Consumption Survey)
A tool developed by the US Energy Information Administration (EIA) to benchmark buildings energy consumption. Www.eia.gov/emeu/cbecs/

CBL (Customer Baseline)
The level of energy usage a customer would have consumed without a demand response program in place. ISOs (Independent System Operators), RTOs (Regional Transmission Organizations) and utilities today do not have a standard CBL methodology to measure and verify the load impacts of demand-response resources.

CBM (Condition-Based Maintenance)
Equipment maintenance that is based on an asset's condition, which is monitored on a periodic or continuous basis for failure attributes, such as vibration or temperature. It is an alternative to failure-based maintenance, initiated when something breaks, or use-based maintenance.

CC (Common Criteria)
An international standard for computer security certification or validation. Under Common Criteria vendors of security systems and products specify functional and assurance requirements and then implement and make claims about the security of their products, which are evaluated by licensed testing laboratories to determine the veracity of the claims.

CCA (Community Choice Aggregation)
An energy policy that can promote distributed and/or renewables generation through community-based contracts with electricity suppliers. The community acts as an aggregator, and residents within it are automatically part of that CCA unless they opt-out, which serves to continue the customer/supplier relationship with the regional IOU (Investor Owned Utilities). The regional IOU is still responsible for delivering power to the CCA members. This policy is available in several states, including California, New Jersey, Massachusetts, Ohio, and Rhode Island.

CCET (Center for the Commercialization of Electric Technologies)
A group of Texas electric and high-tech companies and universities working to enhance the safety, reliability, security, and efficiency of that state's

electric T&D (Transmission and Distribution) system through research, development, and commercialization of emerging technologies. One important objective is to develop and capture the benefits of advancing technologies in transmission, distribution, and end use through pilots and testing programs. (http://www.electrictechnologycenter.com/)

CCF (100 cubic feet)
A standard measurement for water or natural gas consumption. The first "C" is the Roman numeral for 100. 1 CCF equals 748 gallons of water.

CCHP (Combined Cooling, Heat, and Power)
Also known as trigeneration, this is a plant designed to simultaneously generate electricity and useful heating or cooling from a single energy source.

CCMC (CEN-CENELEC Management Centre)
European organization in charge of the daily operations, coordination and promotion of all CEN activities with CENELEC, the standards body focused on electricity.
(http://www.cen.eu/cen/AboutUs/CMC/Pages/default.aspx)

CCPI (Center for Carbon-Free Power Integration)
A new organization established by the University of Delaware that conducts research and facilitates partnerships related to power production from renewable energy sources at the utility level, changes in transmission planning for renewable energy sources, and storage for renewable energy sources. The Center for Carbon-Free Power Integration is part of the University of Delaware Energy Institute and is administered by the UD College of Marine and Earth Studies. (http://www.carbonfree.udel.edu/)

CCRA (Common Criteria Recognition Arrangement)
A collection of international governments organized to ensure that security evaluations of IT products and protection profiles are performed to high

and consistent standards to improve availability of these tested products, eliminate duplicative testing, and improve the evaluation and certification or validation processes for these products. (http://www.commoncriteriaportal.org/ccra/)

CCS (Carbon Capture and Storage (or Sequestration))
Technologies and processes that capture and store carbon dioxide in deep geological formations.

CCSA (China Communications Standards Association)
An industry association focused on standards development for ICT in China. (http://www.ccsa.org.cn/english/about.php)

CDFI (Cable Diagnostic Focused Initiative)
DOE (Department of Energy)-supported research led by NEETRAC to deliver improved diagnostic tools, methods and understanding of cable systems.

CDMA (Code Division Multiple Access)
A proprietary network technology used in 2G and 3G wireless communications. It multiplexes numerous signals through code assignment and uses spread spectrum technology to spread signals over the widest range of channels. It is used in the 800-MHz and 1.9-GHz ranges. CDMA supports nationwide roaming.

CDMA450 (Code Division Multiple Access 450)
Technologies that operate in the 410-740 MHz frequency band in use in 3G networks. Applications include voice and fixed or mobile broadband services.

CDMA2000® (Code Division Multiple Access 2000)
A 3G standards family based on CDMA technology that provides voice and broadband data services over wireless networks. It includes 1Xfor voice services, and EV-DO (Evolution-Data Optimized) broadband data services.

CEA (Canadian Electricity Association)
An organization of Canadian utilities, independent power producers, industry consultants, and manufacturers that works to ensure a safe, secure, reliable, sustainable, and competitively priced supply of electricity. (http://www.canelect.ca/en/home.html)

CEA-2045
The Modular Communications Interface for Energy Management Is a standard developed by the CEA (Consumer Electronics Association) to enable communications for appliances and equipment commonly found in residential dwellings. The interface provides a physical connection from a communication module to devices and a communications protocol with OSI (Open System Interconnection) layer specifications including application layer messaging. An optional translation function is specified for connection to another communications medium.

CEA-852.1:2009
This is the standard that defines tunneling, a communications method that allows data acquisition and control devices to communicate via the Internet. It is a part of the LonWorks suite of standards.

CEATI (Centre for Energy Advancement through Technological Innovation International)
An international technology solutions exchange and development program for over 120 utilities and research organizations. Interest groups are organized around generation, transmission, and distribution topics. The program model is built to combine inter-utility information exchange and informal benchmarking with practical projects of immediate impact. It was historically associated with CEA. (http://www.ceati.com/index.php)

CEC (California Energy Commission)
The primary energy planning and policy entity for the State of California. Its major responsibilities include forecasting future energy needs, setting the

state's appliance and building efficiency standards, supporting energy research, development, and demonstration programs, and providing market support to existing, new, and emerging renewable technologies. The CEC manages the PIER (Public Interest Energy Research) program, which is responsible for public and private Smart Grid RD&D programs for transmission, distribution, distributed generation, and consumer technologies. The Energy Commission awards funding in electricity-related RD&D, and natural gas RD&D. (www.energy.ca.gov/)

CEC IEPR (California Energy Commission Integrated Energy Policy Report) A biennial integrated energy policy report prepared in response to California SB 1389 with an assessment of major energy trends and issues facing the state's electricity, natural gas, and transportation fuel sectors. It provides policy recommendations to conserve resources; protect the environment; ensure reliable, secure, and diverse energy supplies; enhance the state's economy; and protect public health and safety. It proposes policy and program direction and specific recommendations to meet energy policy goals. (http://www.energy.ca.gov/2009publications/CEC-100-2009-003/CEC-100-2009-003-CMF.PDF)

CECORA (CEllular Infrastructure: COgnitive RAdio Networks Beyond 4G) An EU-based group of industry and academic organizations focused on advancing cellular communications through the use of CRN technology in 4G networks.

CEE (Consortium of Energy Efficiency) A nonprofit organization that develops initiatives for its members to promote the manufacture, purchase, and adoption of energy-efficient products and services. CEE members include utilities, statewide and regional administrators, environmental groups, research organizations, state and provincial energy offices in the USA and Canada, manufacturers, and retailers. Both DOE (Department of Energy) and EPA (Environmental Protection Agency) support and fund this organization. (http://www.cee1.org/)

CEEP (Clean and Efficient Energy Program)
A partnership of APPA (American Public Power Association), LPPC (Large Public Power Council), and ASE (Alliance to Save Energy) that is a nationwide initiative to assist public power utilities in the planning, design, implementation, and evaluation of energy efficiency and renewable energy activities. (http://cleanefficientenergy.org/)

CEER (Council of European Energy Regulators)
A nonprofit organization that facilitates the creation of a single, competitive, efficient and sustainable internal market for gas and electricity in Europe. It enables information exchange and assistance between national energy regulators and is the European interface with the European Commission, and the Directorate General Transport and Energy, Directorate General Competition and Directorate General for Research. It works to ensure consistent application of competition law to the energy industry. It also shares regulatory experiences with the North American Regulatory Commissioners Association (NARUC), ERRA (European Energy Research Alliance) and the International Energy Regulation Network (IERN). (http://www.energy-regulators.eu/portal/page/portal/EER_HOME/EER_ABOUT/CEER)

CEFACT (Centre for Trade Facilitation and Electronic Business)
A UN organization that encourages close collaboration between governments and private business to secure interoperability for the exchange of information between the public and private sector. (http://www.unece.org/cefact/about.htm)

CEIC (Carnegie Mellon Electricity Industry Center)
An educational center of excellence established by the Sloan Foundation. Its primary mission is to work with industry, government, and other stakeholders to address the strategic problems of the electricity industry. Funding primarily comes from the Sloan Foundation, EPRI (Electric Power Research Institute), US National Science Foundation, Environmental

Protection Agency, DOE (Department of Energy), and governmental organizations. (http://wpweb2.tepper.cmu.edu/ceic/)

CEMI$_N$ (Customers Experiencing Multiple Interruptions)
A reliability metric that measures the percentage of customers that experience more than N sustained interruptions in a year's time. N is the threshold number that is defined by each utility or regulatory agency.

CEN (European Committee for Standardization)
A European standards and technical specifications organization that works with electrotechnology (CENELEC) and telecommunication (ETSI) organizations.

CENELEC (European Committee for Electrotechnical Standardization)
A nonprofit technical organization comprised of the National Electrotechnical Committees of 31 European countries. It develops voluntary electrotechnical standards that help develop the Single European Market/European Economic Area for electrical and electronic goods and services. (http://www.cenelec.eu/Cenelec/Homepage.htm)

Centre for White Space Communications
An R&D center comprised of industry and academic entities that is focused on the engineering and technological issues associated with White Space spectrum and cognitive radio. (http://www.wirelesswhitespace.org/)

CEOs for Cities
A nonprofit organization made up of cross-sector urban leaders whose goal is to make cities more connected, innovative, and talented. Work projects include research on vital city infrastructure and performance, city dividends to measure how a city operates and where it can improve, as well as programs to reduce pollution and increase opportunity. (www.ceosforcities.org)

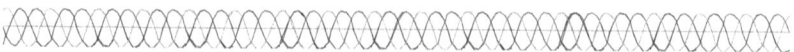

CEPT (European Conference of Postal and Telecommunications administrations)
An association of EU governmental agencies that focus on commercial, operational, regulatory and technical standardization issues through three autonomous committees: the Electronic Communications Committee (ECC), the European Committee for Postal Regulation (CERP), and the Committee for ITU Policy (Com-ITU). Of these three committees, the ECC and the Com-ITU have influence in Smart Grid and M2M communications. (http://www.cept.org/)

Cermet (Ceramic -Metal)
Lithium composites developed for improved performance of lithium ion batteries that boosts battery power by increasing the available surface area.

CERP-IoT (Cluster of European Research Projects on the Internet of Things)
An initiative to aggregate European research projects together to define and promote a common vision of the Internet of Things including RFID. Research projects include privacy concerns. It is part of the FP7 initiative. (http://www.rfid-in-action.eu/cerp)

CERTS (Consortium for Electric Reliability Technology Solutions)
An organization sponsored by the DOE (Department of Energy) and the CEC (California Energy Commission) to research, develop, and disseminate new methods, tools, and technologies to protect and enhance the reliability of the USA electric power system and efficiency of competitive electricity markets. Research covers five areas: realtime grid reliability management, reliability and markets, distributed energy resources integration, load economics, and grid reliability technology gaps. CERTS is engaged in research on microgrids. (http://certs.lbl.gov/)

CES (Community Energy Storage)
Small battery energy storage systems that are located in the distribution system and help to stabilize electricity as well as offer storage options for distributed generation by capturing excess energy produced by homes and re-dispatched later. These systems are generally around 25 KW and provide 1 to 2 hours of backup power for small groups of residential users.

CESA (California Energy Storage Alliance)
An energy storage advocacy association for California. It is a technology-neutral association of companies and organizations focused on the rapid expansion of energy storage to promote growth of renewable energy and a more reliable and secure electric system. (http://www.storagealliance.org/)

CFTC (United States Commodities Future Trading Commission)
An independent agency mandated to protect market users and the public from fraud and abuse related to the sale of commodity and financial futures and options. The Dodd-Frank Act of 2009 added energy futures, including electricity, to this agency's oversight. (www.cftc.gov)

CHAdeMO Association (Charge De Move)
An industry association comprised of Japanese car manufacturers and electric utilities promoting electric vehicles through the efforts of technical improvements of quick chargers, standardization activities of charging methods, and international extension of knowledge related to quick-charger installations. (http://www.chademo.com/indexa.html)

Charger
A term used in the automotive industry to identify an EV (Electric Vehicle) or PHEV (Plugin Hybrid Electric Vehicle) device that converts AC to DC power for auto battery use.

Charging infrastructure
The network of wiring, charging stations, and distribution grid components to support charging/discharging for EV, PHEV, or any other electrified contraptions for personal or public transport.

Charging Level 1
See Level 1 Charge.

Charging Level 2
See Level 2 Charge.

Charging Level 3
See Level 3 Charge.

Charging station
Equipment that supports the charging/discharging of any type of electrified personal or public transport.

Check valve
A valve that prevents any reversal in the flow in the system, controlling the flow direction.

CHP (Combined Heat and Power)
A plant designed to produce both electricity and useful heat from a single heat source.

CHP Partnership (Combined Heat and Power Partnership)
A voluntary program within the Environmental Protection Agency that works to reduce the environmental impact of power generation through CHP technology. The Partnership works with energy users, the CHP industry, state and local governments, and other entities to develop CHP projects and promote the environmental and economic benefits of CHP. (http://www.epa.gov/chp/index.html)

CIA (Confidentiality, Integrity, Availability)
A security industry acronym for three important objectives of cyber security design and deployment. Confidentiality ensures that only the intended recipients receive and are able to interpret the transmitted data. Integrity concerns authorized access to modify sensitive data. Availability focuses on granting access to authorized users or viewers of that sensitive data. The priority of these three objectives may change based on where the data resides – close to a network edge or within several layers of network security.

CICC (Critical Infrastructure Communications Coalition)
An industry association advocating for regulatory policies that promote the ICT capabilities needed to protect and maintain the nation's critical infrastructures such as electrical, gas, and water grids.
http://www.utc.org/utc/critical-infrastructure-communications-coalition

CIGRE (International Council on Large Electric Systems)
A global, nonprofit organization that researches and makes recommendations on topics including planning and operation of power systems; design, construction, maintenance, and disposal of high-voltage equipment; and issues related to protection of power systems, telecontrol, telecommunication equipment, and information systems. Its recommendations are used in standards development by the IEC and other international organizations. (http://www.cigre.org/)

CIKR (Critical Infrastructure and Key Resource)
A term used by the Department of Homeland Security and other security agencies to designate critical infrastructure (the most important assets, systems, and networks) and key resources (the essential public or private assets necessary for economic or governmental operations). These designations help to assess risks, vulnerabilities, and threats in order to develop mitigation plans.

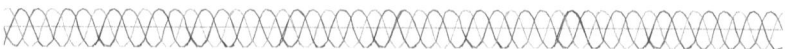

CIM (Common Information Model)
An abstract model of utility-specific data that may be used as a reference, a category scheme of data names and applied meanings of data definitions, database design, and a definition of the structure and vocabulary of message schemas. The CIM also includes a set of services for exchanging data, called GID (Generic Interface Definition). CIM was adopted by IEC (International Electrotechnical Commission) working groups for use in standards development.

CIMug (CIM Users Group)
A technical subcommittee of UCAIug that is responsible for handling all technical and maintenance issues concerning the CIM and related standards and developing consensus and consistency across the utilities industries. (http://cimug.ucaiug.org/default.aspx)

CIP (Critical Infrastructure Protection)
NERC (North American Electric Reliability Corp.) standards established to ensure cyber security of utility assets deemed as critical assets in the bulk power system. CIP-002-1 through CIP-011-1 and CIP-014-1 are the specific standards focused on cyber security and the definitions of critical assets, cyber assets, and critical cyber assets. FERC (Federal Energy Regulatory Commission) Order 706 approved these security standards: 002- Critical Cyber Asset Identification, 003- Electronic Security Perimeter, 004- Personnel and Training, 005- Electronic Security Perimeter, 006- Physical Security of Critical Cyber Assets, 007- Systems Security Management, 008- Incident Reporting and Response Planning, and 009- Recovery Plans for Critical Cyber Assets. 010 defines Cyber System Categorization, 011 covers Cyber System Protection. 014 is the most recently added, and addresses the Physical Security of Transmission Stations and Substations.

CIPC (Critical Infrastructure Protection Committee)
The committee that coordinates NERC's (North American Electric Reliability Corp.) security initiatives. It comprises industry experts in the areas of cyber

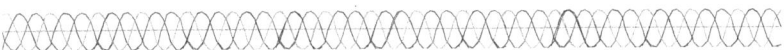

security, physical security, and operational security. CIPC reports to NERC's Board of Trustees. It is governed by an Executive Committee, which manages CIPC policy matters and provides support to CIPC's subcommittees and their working groups and task forces. This committee collaborates with the DOE (Department of Energy), the Department of Homeland Security, and Public Safety and Emergency Preparedness Canada on critical infrastructure and security matters. (http://www.nerc.com/page.php?cid=1|9|117|139)

CIPP (Certified Information Privacy Professional)
A professional certification in information privacy offered by IAPP (International Association of Privacy Professionals) that can be distinguished by region (US, Canada, or Europe) or function (government or Information Technology (IT)).

Circuit
A conductor or a system of conductors through which electric current flows.

Circuit-mile
The total length in miles of separate circuits regardless of the number of conductors used per circuit.

CIS (Customer Information System)
A software application that contains customer information and functions to perform customer service, billing, accounting, and other automated operations.

CISA (Certified Information Systems Auditor)
A professional certification for information systems audit, control and security skills that is offered by ISACA (Information Systems Audit and Controls Association).

CISM (Certified Information Security Manager)
A professional certification for individuals who design, assess, or manage enterprise information security that is offered by ISACA (Information Systems Audit and Controls Association).

CISO (Chief Information Security Officer)
The highest ranking executive responsible for information security in an organization.

CISSP® (Certified Information Systems Security Professional)
A global information security certification managed by the International Information Systems Security Certification Consortium or (ISC)². The credential is accredited by ANSI ISO/IEC Standard 17024:2003. The CISSP has been adopted as a baseline for the USA National Security Agency's ISSEP program.

Cistern
An above- or below-ground storage tank that collects rainwater runoff.

Clarification
A process in surface water treatment plants that involves removing particles such as dirt or organic matter from water, usually by adding chemicals that alter particle charges to settle or float them out of water.

Class rate schedule
An electric rate schedule applicable to one or more specified classes of service, groups of businesses, or customer uses.

Classes of service
A class or group of customers with similar characteristics that have a common rate for electric service. Some common classifications are residential, commercial, industrial, and transportation.

Clean Energy Jobs Plan

A California state initiative that calls for the installation of 20,000 MW of renewable energy resources by 2020, of which 12,000 MW should be localized energy resources or LER; increased energy efficiency in commercial and residential buildings and appliances; encourage deployment of combined heat and power systems and energy storage systems; and streamline processes to approve renewable energy deployments. (http://gov.ca.gov/docs/Clean_Energy_Plan.pdf)

Clean Energy Technology

As defined in the ARRA Section 1306 amending EISA 2007 (Energy Independence and Security Act of 2007), technology related to the production, use, transmission, storage, control, or conservation of energy that improves the energy efficiency of any fuel, improves and extends sustainable practices, or has an end benefit of stabilizing and reducing GHG (Greenhouse Gas) emissions.

Clean Water Act

See CWA.

Cleantech

Any technology, process, or service that reduces or eliminates GHG or any other pollution outputs, reduces waste, or enhances sustainability practices.

ClimateTalk Alliance

An industry association focused on interoperability through a common communication infrastructure for HVAC and Smart Grid devices. It has three steering committees and five working groups focused on wired and wireless communications and applications. (http://www.climatetalk.com/)

Cloud Security Alliance
A not-for-profit organization of industry practitioners, associations, vendors and others promoting the use of best practices for providing security assurance for cloud computing. (https://cloudsecurityalliance.org)

CLSF (Carbon Sequestration Leadership Forum)
An international climate change initiative promoting carbon capture and storage technologies. Members include national governmental entities that produce or use fossil fuels and are willing to commit R&D to development of cost-effective technologies and processes for capture and storage. (http://www.cslforum.org/)

CME (Canadian Manufacturers and Exporters)
A nonprofit association of companies (including those manufacturing Smart Grid technologies) promoting the competitiveness of Canadian manufacturers and exporters in markets around the world. (www.cme-mec.ca)

CMIP/CMIS (Common Management Information Services/Common Management Information Protocol)
These are OSI (Open Systems Interconnection) implemented protocols and services defined in ISO/IEC 9595 and 9596 respectively. They are used to manage available network services, gather information available on a given network, and provide security in the form of access control. Combined, they manage and enable functionality at all levels of the OSI model.

CMO (Change Meter work Order)
A transaction to make a change on an electric meter.

CMOM (Change Meter Order Management)
The systems and processes to schedule and dispatch meter work orders.

Code of Federal Regulations
A compilation of the general and permanent rules of the executive departments and agencies of the federal government as published in the Federal Register. The code is divided into 50 titles that represent broad areas subject to federal regulation. Title 18 contains the FERC (Federal Energy Regulatory Commission) regulations.

Cogeneration
Production of electricity and useful thermal energy from a single fuel source, typically located at or near the point of consumption.

COGEN Europe
A European association for the promotion of cogeneration and CHP (Combined Heat and Power). Its primary goal is to work towards widespread adoption of CHP technologies in a variety of buildings and uses. (www.cogeneurope.eu)

Cognitive radio
See CR.

Coincident resources
The amount of DR (Demand Response) curtailments, or energy saved, if all DR products simultaneously reduced load by the quantity specified in DR programs. All enrolled customers are assumed to respond as outlined in DR program contracts.

COM-ITU (Committee – International Telecommunications Union)
A CEPT (European Conference of Postal and Telecommunications administrations) committee that has influence in Smart Grid and M2M communications. It coordinates CEPT activities with the ITU. (http://www.cept.org/com-itu)

Combined cycle plant
A type of gas turbine generator that produces electricity from otherwise lost waste heat or exhaust gas. This heat is routed to a boiler or heat recovery steam generator and the hot steam is expanded in a steam turbine to produce electricity. It increases the efficiency of the electric power generation.

Combined household energy expenditures
The total amount of funds spent for energy consumed in, or delivered to, a housing unit during a given period of time and for fuel used to operate the motor vehicles that are owned or used on a regular basis by the household. The total dollar amount for energy consumed in a housing unit includes state and local taxes, but excludes merchandise repairs or special service charges.

Combined hydroelectric plant
A hydroelectric plant that uses both pumped water and natural streamflow for the production of power.

Combined pumped-storage plant
A pumped-storage hydroelectric power plant that uses both pumped water and natural streamflow to produce electricity.

Combined sewer system
A system that carries stormwater runoff in addition to wastewater from residential, commercial, and industrial uses.

Combustion
Chemical oxidation accompanied by the generation of light and heat.

Combustion chamber
An enclosed vessel in which chemical oxidation of fuel occurs.

Commercial building

A building with more than 50 percent of its floor space used for commercial activities. Government buildings are included, except those on military bases or reservations.

Commercial facility

An economic unit that is owned or operated by one person or organization and occupies two or more commercial buildings at a single location. A university and a large hospital complex are examples of a commercial multi-building facility.

Commercial water use

Water used for commercial facilities and institutions that is sourced from public-supplied sources and self-supplied sources, such as a well.

Commissioning

A sequence of actions undertaken when equipment such as refrigeration or facility EMS systems are installed that includes sizing and testing for particular duty cycles and operations based on occupancy, weather, and other factors. As factors change, this equipment must be re-commissioned or retro-commissioned. Smart Grid technology could enable continuous commissioning to increase energy efficiency based on continuous monitoring and controlling functions.

Common Criteria (CC)

An international standard for computer security certification or validation. Under Common Criteria vendors of security systems and products specify functional and assurance requirements and then implement and make claims about the security of their products, which are evaluated by licensed testing laboratories to determine the veracity of the claims.

Common Criteria Recognition Arrangement (CCRA)
Recognition that IT products and protection profiles which earn a Common Criteria certificate do not require further evaluation of their security claims for purchase or use.

Community-based storage
A practice of integrating 1MW to 10MW energy storage into subdivisions and towns for load balancing and frequency regulation.

Community Water System
Facilities that are owned by a mix of public owners, private owners and cooperatives. The Bioterrorism Act defines small community water systems as water systems as those serving a population of greater than 3,300 but less than 50,000; medium-sized water systems as those serving more than 50,000 people but less than 100,000; and large systems as those serving population of more than 100,000 people.

Compaction
Transmission line designs that deliver more power flowing through line rights of way. Compaction reduces utilities' capital and maintenance costs and line losses.

Competitive merchant generators
Companies that build electric power capacity on a speculative basis or have acquired utility-divested plants. These companies then market their output at competitive rates in unregulated markets.

COMTRADE (COmmon format for TRAnsient Data Exchange for power systems)
An IEEE standard describing a file format used in substations for oscilloscope data for data storage. It is also becoming the default file format for phasor data. It does not cover transmission of data.

Conductor
Metal wires, cables, and bus-bar used for carrying electric current.

Congestion
An insufficient transfer capacity to simultaneously supply all orders for electricity transmission. Congestion may be a limitation in physical capacity of transmission lines or caused by operational restrictions deployed to protect grid security and reliability.

Connected Lighting Alliance
A nonprofit group that advocates the use of interoperable and standardized wireless lighting control technologies.

Connected load
The sum of the continuous ratings or the capacities for a system, part of a system, or a customer's consumption.

Connection
The physical connection between two electric systems to transfer electricity.

Conservation Voltage Reduction
See CVR.

Constant dollar analysis
An economic analysis used in comparing technologies to recognize the potential for advancement of a component or solution through R&D for improved performance and cost reductions. Constant-dollar analysis does not incorporate inflation effects in capital carrying charges and operating cost projections. Constant-dollar analysis gives a clearer picture of real cost trends and purchasing power differences.

Constituent
A measurable chemical or biological substance in water or sediment.

Constrained facility
A transmission asset such as a line or transformer that is near or at its SOL (System Operating Limit) or (IROL) Interconnection Reliability Operating Limit.

Consumer charge
An amount charged periodically to a consumer for utility costs, such as billing and meter reading that is not based on energy consumption.

Consumption
Use of electrical energy, typically measured in kilowatt hours or kWh.

Consumptive use
Water that is removed from the water environment and not available for reuse. Also known as water consumption.

Contingency
The unplanned failure of system components, such as generators, transmission lines, or other electrical assets.

Contingency reserve
A generating system capacity held in reserve to cover any failure of system components, such as generators or transmission lines.

Continua Health Alliance
An industry organization of technology, medical device, and healthcare companies collaborating to improve the quality of personal telehealthcare through interoperable sensors, home networks, and applications that support health solutions from a variety of vendors. This is an extension of

the connected home into health and well-being applications. (www.continuaalliance.org)

Continuous commissioning
A combination of processes, hardware, and software to ensure that buildings are operating at peak energy efficiency to reduce overall energy costs and carbon emissions and optimize performance of building HVAC gear.

Continuous delivery energy sources
Energy sources provided continuously to a building.

Control Area
The ISO (Independent System Operator) territory where the ISO is responsible for scheduling generation and load, contracting for grid reliability (ancillary) services, and scheduling imports and exports of electricity. These areas have detailed protocols or guidelines to follow, which have impacts on Smart Grid projects.

Control valve
A valve plus an actuator that can appear as a ball valve, butterfly valve or other valve type.

Conversion efficiency
The amount of energy that can convert to electricity compared with the potential energy of that renewable energy source.

Conveyance loss
Water lost from pipes, canals or ditches by leakage or evaporation.

Cooling degree days
A method of measuring over a time period the warmth of a location compared with a base temperature, usually 65 degrees Fahrenheit. To compute, subtract the base temperature from the daily average (high and

low temperatures – set negative temperatures to zero). Sum up each daily calculation to create a measurement for a particular time period. This measurement is used in energy analysis to develop AC energy requirements.

Cooling tower
Power plant facility that transfers heat in cooling water to the atmosphere either by direct evaporation or by convection and conduction.

Cooperative electric utility
An electric utility owned and operated by its energy consumers. The cooperative generates, transmits, and/or distributes electrical energy to a specific area. These utilities are usually exempt from federal income tax laws and were initially financed by the Rural Utilities Service of the USA Department of Agriculture (formerly the Rural Electrification Administration).

Cooperative water utility
A water utility owned and operated by its consumers. The cooperative sources and distributes water in a specific region – usually in rural areas and for small towns.

Coordination service
An exchange of electricity between two or more utilities that generally have sufficient capacity to provide for their normal load requirements.

COP (Coefficient of Performance)
A measure of the energy efficiency of heating and cooling devices. It is calculated as the heating or cooling capacity in Btu/h divided by electrical output in Btu/h. The higher the COP, the more efficient the device.

CoRE (Constrained Restful Environment)
An IETF working group developing a framework for applications running on IP networks that have constraints such as limited packet sizes, high packet loss, or devices that only periodically power up for brief periods of time.

COSEM (Companion Specification for Energy Metering)
An object-oriented interface model of communicating energy metering equipment, providing a view of the functionality available through the communication interfaces. Meters have a worldwide unique identifier called the logical device name. The COSEM model sets the rules for data exchanges any energy meter in a manufacturer independent, controlled, and secure way.

Cost of service
A rate-making framework that lets utilities recover the cost of providing electrical services.

Cost of service regulation
A pricing mechanism that lets the utility set rates based on the cost of providing services to customers and the right to earn a limited profit.

COTS (Commercial Off The Shelf)
Shorthand description of creating a solution from available hardware and software without the need for customized or new development.

Coupler
A device that transfers the communications signal to and from medium-voltage and low-voltage power lines in a BPL (Broadband Power Line) or PLC (Power Line Carrier) network, and also delivers a real-time read of the current of the electrical distribution grid.

CPP (Critical Peak Pricing)
A pricing option or rate structure often used by utilities as part of a demand response program. The objective is to reduce energy use during peak periods with customers who receive discounted pricing in exchange for agreeing to reduce their energy demand during those utility-identified peak periods. It is a form of dynamic pricing. It is classified by the NERC (North American Electric Reliability Corporation) as non-dispatchable DR (Demand Response).

CPP with Controls (Critical Peak Pricing with Controls)
A demand response management that combines direct remote control of consumer appliances like air conditioning with a high price for use during specified critical peak periods. It is triggered by system contingencies or market prices and is a form of dynamic pricing. It is classified by NERC (North American Electric Reliability Corporation) as a form of dispatchable DR (Demand Response).

CPUC (California Public Utilities Commission)
The commission that regulates investor-owned electric, natural gas and water companies within California. It consists of five governor-appointed commissioners and staff to ensure that consumers have safe, reliable utility service at reasonable rates, protect against fraud, and promote the health of California's economy. (http://www.cpuc.ca.gov/puc/)

CPV (Concentrated PhotoVoltaics)
A technology that focuses lighting using mirrors to a smaller PV module. Categorized as low, medium, or high concentration, it includes lenses and mirrors, reflectors, and receivers to generate electricity.

CR (Cognitive Radio)
Software-defined radio technologies such as transceivers with intelligent communication systems that are aware of location and RF frequency spectrum usage, and automatically and dynamically schedule, share, or

change frequencies to make optimal use of available frequency spectrum and wireless networks. It is an enabling technology for M2M communications.

CRADA (Cooperative Research and Development Agreement)
An agreement between the federal government and private sector participants to work together on a mutually beneficial project. It is a way to combine private-sector funding, resources, and facilities with the federal research labs.

C-RAS (Centralized Remedial Action Scheme)
A SIPS (System Integrity Protection Scheme) that uses IEDs (Intelligent Electronic Devices) to monitor current grid conditions, detect events, and communicate with a central control processor to determine if generation or load must be remotely dropped.

CREZ (Competitive Renewable Energy Zone)
Geographic areas throughout a state in which renewable energy resources and suitable land areas are sufficient to develop generating capacity from renewable energy technologies.

Critical Assets
A NERC (North American Electric Reliability Corporation) definition specific to Critical Infrastructure Protection requirements. It covers facilities, systems, and equipment that, if destroyed, degraded, or otherwise rendered unavailable, would affect the reliability or operability of the bulk electric system.

Critical Cyber Assets
A NERC (North American Electric Reliability Corporation) definition specific to Critical Infrastructure Protection requirements. These are Cyber Assets that are essential to the reliable operation of Critical Assets.

CRN (Cognitive Radio Networks)
An emerging concept of using underutilized licensed spectrum for selected communications that typically reside in unlicensed spectrum. It is considered a promising approach to enable communications in the fully deployed IoT.

CRN® (Cooperative Research Network)
An NRECA (National Rural Electric Cooperative Association) service that monitors, evaluates, and applies technologies that help electric co-ops control costs, improve productivity, and enhance member and customer service. Six focus areas exist, and each has a members-based advisory group that develops the research agendas and monitors projects. Energy efficiency, energy storage technologies, and evaluation of broadband applications are some topics within the focus areas.

Cross-connection
An unprotected actual or potential connection between a system used to supply drinking water and any source or system containing unapproved water or a substance that is not or cannot be approved as safe, wholesome, and potable.

CRR (Congestion Revenue Rights)
Insurance-like financial products designed to reduce transmission congestion costs. These hedging tools provide the holder reimbursement of the congestion charges in the day-ahead market and thereby provide transmission service customers with price certainty.

CRTC (Canadian Radio-television and Telecommunications Commission)
The Canadian federal agency empowered by Parliament to regulate and supervise the broadcasting and telecommunications systems in Canada. (http://www.crtc.gc.ca/)

Cryptography
The science of securing information, generally through the use of encryption algorithms in communications networks and computers.

CS (Cyber Security)
A term for either computer networks security or a working group within NIST (National Institute of Standards and Technology) focused on cross-cutting security issues with grid interoperability. (www.nist.gov/smartgrid/)

CSCC (Communications Sector Coordinating Council)
An industry organization that works with specific USA federal agencies on activities and coordination of initiatives to improve physical and cyber security of critical infrastructures, including Smart Grid networks. (http://www.commscc.org/)

CSEP (Consortium for Smart Energy Profile 2.0 interoperability)
A business forum formed to unify and accelerate SEP 2 interoperability. The Consortium was formed in 2011 by the HomePlug Alliance, Wi-Fi Alliance and ZigBee Alliance. www.csep.org

CSI (California Solar Initiative)
An initiative of the CPUC (California Public Utilities Commission) and the CEC (California Energy Commission) that offers financial incentives for solar installations to achieve an overall goal of 1,750 megawatts of solar by 2016. It is administered by the California IOUs (Investor Owned Utilities): Southern California Edison, Pacific Gas and Electric, and San Diego Gas and Electric (via the California Center for Sustainable Energy). (http://www.gosolarcalifornia.org/csi/index.html)

CSO (Combined Sewer Overflow)
The discharge of untreated sewage and stormwater that exceeds peak flow capacity at water treatment plants that typically occurs during high runoff.

CSP (Concentrating Solar Power)
A collection of technologies that use mirrors to heat liquids to steam to drive turbines. Typical utility-scale plants generate at least 50 MW or more of electricity.

CSP (Curtailment Service Provider)
Demand response providers that are different from utilities or other load-serving entities. CSPs may promote demand response programs and sell the demand response load to utilities, RTOs (Regional Transmission Organizations), and/or ISOs (Independent System Operators).

CSSP (Control Systems Security Program)
A DHS (Department of Homeland Security) program that provides information on cyber threats, including vulnerability and mitigation, to industrial control systems owners, operators and vendors. It coordinates activities to reduce the likelihood of success and severity of impact of a cyber attack against critical infrastructure control systems through risk mitigation activities.

CSWG (Cyber Security Working Group)
A NIST (National Institute of Standards and Technology) group composed of vendors, service providers, academia, regulatory organizations, national research labs, federal agencies, and private citizens focused on developing recommendations for standards in cybersecurity and privacy regarding the Smart Grid. It produced the NISTR 7628.

CTIA (Cellular Telecommunications Industry Association)
An international nonprofit membership organization representing the wireless communications industry in educational and governmental advocacy. Members include wireless carriers and manufacturers of wireless data services and products. (http://www.ctia.org/)

CUPSS (Check Up Program for Small Systems)
An asset management tool for small drinking water and wastewater utilities, provided by the EPA in the USA.

Current
The flow of electricity in an electrical conductor. Current is measured in amperes.

Current dollar analysis
An economic analysis used by utilities that includes the effect of inflation in calculations as the basis for budgeting future expenditures. Current-dollar analysis includes expected effects of inflation on capital carrying charges and operating costs. Current-dollar analysis more closely approximates future cash flows than constant dollar analysis, which is important when utilities are reviewing estimates with regulatory authorities and security analysts.

Current meter
A meter that measures the speed of flowing water.

Curtailment
Load reductions for set periods of time.

CVR (Conservation Voltage Reduction)
Techniques to reduce system voltage levels by a percent or two and achieve overall energy savings without sacrificing power quality. Smart meters may be deployed as sensors to help utilities more granularly modulate voltages throughout distribution grids.

CWA (Clean Water Act)
Federal legislation that created national objectives to restore and maintain the chemical, physical, and biological integrity of the Nation's waters. Initially known as the Federal Water Pollution Control Act of 1948, it was

renamed in 1977. Major amendments were enacted in 1961, 1966, 1970, 1972, 1977, 1987, and 2008 including establishment of uniform water quality standards and a Federal Water Pollution Control Administration to set quality standards when states fail to act. Provisions include a requirement that the Federal Power Commission not grant a license for a hydroelectric power project to regulate streamflow for the purpose of water quality unless certain conditions are satisfied. 1977 changes required localities to consider alternatives to the traditional centralized treatment facilities, resulting in investments in decentralized treatment systems. It was initially known as the Water Pollution Control Act of 1948.

CWRA (Canadian Water Resources Association)
An organization committed to building awareness of the value of water and promoting responsible and effective water resource management in Canada. Members include water users and water resource professionals, scientists, and academics. (http://www.cwra.org/)

CWWA (Canadian Water and Wastewater Association)
A nonprofit organization representing Canada's public sector municipal water and wastewater services and private sector suppliers and partners with respect to policies, programs, national codes, standards, and legislation. (http://www.cwwa.ca/)

Cyber Assets
A NERC (North American Electric Reliability Corporation) definition specific to Critical Infrastructure Protection requirements. These are programmable electronic devices and communication networks including hardware, software, and data that are used in the bulk electric system.

Cyber security incident
Any event that tries to or achieves compromise of the electronic or physical security perimeter of a critical cyber asset or disruption of the operation of a critical cyber asset.

D

DADS (Demand Response Availability Data System)
An in-development data collection system for DR (Demand Response),
designed by NERC's (North American Electric Reliability Corporation) DR
Data task force and NAESB's (North American Energy Standards Board)
M&V (Measurement and Verification) standards committee.

DAS (Distributed Antenna System)
A deployment of small antennas in indoor or outdoor settings connected to
a base station. DAS systems improve radio frequency (RF) coverage in
challenging environments.

DAS (Distribution Automation System)
As defined in the ARRA Section 1306 that amends EISA 2007 (Energy
Independence and Security Act of 2007), a system that operates at or
below the voltage level of a distribution substation that enables realtime,
remote monitoring and control of components in the distribution grid. It
increases operational efficiency and reduces losses in the distribution grid
through capabilities such as remote outage detection and restoration,
power loss detection and prevention, and predictive maintenance and
diagnostics.

DAS Forum (Distributed Antenna System Forum)
An industry forum promoting micro and pico cell deployments to
supplement macro cell deployments in wireless networks.
(http://www.thedasforum.org/)

DASH7
A wireless sensor networking technology that operates at 433 MHz, a
globally available, unlicensed spectrum. It is based on the ISO 18000-7
standard and is characterized by low power requirements, penetration of
walls and other spectral barriers, and 2-km range.

DASH7 Alliance
An industry association promoting the Dash7 spectrum. It provides a framework for application development, interoperability, and security for DASH7-enabled transactions. It provides a framework for extensive application development, seamless interoperability, and security for DASH7-enabled transactions. (http://www.dash7.org/)

Data broker
An entity that collects data from a variety of sources, including Internet and online sources as well as databases, print documentation, and surveys to package and sell as a product or service to other entities. It can include personal consumer data or business data to serve information needs of private-sector and governmental agencies, and could include information generated by Smart Grid or M2M applications. Also known as an information broker.

Data center
Any facility that stores computers and communications equipment to process, store, and transmit digital information.

Data privacy
A concept of protection and control of an individual's data boundaries, much like physical privacy presumes protection against intrusions. Data privacy includes personal identifiers like name and address; "how you live" data such as behavioral or energy consumption data or health and medical data; and locational data – where you are at any given moment in time. Data privacy involves not only protecting access to that data, but also allowing the associated individuals to have control over how that data is used, shared, stored and destroyed. Data privacy also encompasses the analysis of data items, each of which may not be personally identifiable, but as a group may reveal personal activities or other aspects of the associated individuals' lives. The data privacy concept is framed in legislation by many

countries to protect and balance individual expectations or rights to privacy against legitimate needs for other entities to have access to that data.

Day-ahead and hour-ahead markets
Forward-looking markets in which electricity quantities and market-clearing prices are calculated individually for each hour of the day on the basis of bids for electricity sales and purchases.

Day-ahead schedule
A schedule prepared by a scheduling coordinator or the independent system operator before the beginning of a trading day. This schedule indicates the levels of generation and demand scheduled for each settlement period in that trading day.

dB (Decibel)
A logarithmic unit used to describe a ratio, such as power or voltage. It can be used to measure transmission loss or gain in communications networks.

dBm
The power level in decibels relative to 1 MW. It is used in transmission grids as a measure of absolute power.

DC (Direct Current)
Electricity that flows continuously in the same direction. DC is converted to AC for use in a typical 120-volt or 220-volt appliances. DC is used in industrial applications and appliances that use battery power.

DCIE (Data Center Infrastructure Efficiency)
A metric used to determine the energy efficiency of a data center. It is calculated by dividing IT equipment power by total facility power.

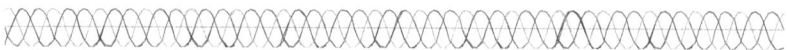

DCLG (Department for Communities and Local Government)
A department within the Government of the United Kingdom that is responsible for housing planning and development. A strong focus of DCLG is ensuring new building construction and developments are built with energy usage reduction technologies.

DDoS (Distributed Denial of Service)
A coordinated Internet attack that is conducted from a large number of locations across a network to disable a website. These attacking locations are sometimes hacked systems, and their owners may be unaware of the involvement of their computers in the attack.

DDS (Data Distribution Service)
This is an OMB (Object Management Group) M2M standard that enables real-time, interoperable data exchanges. It is an industrial internet protocol that enables network interoperability for connected machines, enterprise systems and mobile devices.

DE (Decentralized Energy)
Electricity production at or near the point of use, regardless of size, technology, or fuel, used for both off-grid and on-grid use. It is sometimes known as distributed generation, although distributed generation usually implies connection to the grid.

Decoupling
A regulatory and market strategy that allows utilities to invest in and profit from efficiency-based capacity by assuring them a return that is equivalent to sales of electricity. This policy decouples utility fixed-costs recovery from electricity sales. Utilities collect revenues based on the amount determined by their local regulatory agencies, usually calculated on a per-customer basis. Periodically, revenues are reviewed for rate adjustments to ensure the pre-determined revenue requirement. This strategy is deployed in 17

states at the current time with several other states in the process of setting up utility mechanisms to support decoupling.

DECT (Digital Enhanced Cordless Telecommunications)
A communications standard that may be repurposed if it is deployed in battery-powered, web-connected home automation applications that are controlled by smart phone apps. DECT uses the 1.9 GHz frequency, and uses ETSI (European Telecommunications Standards Institute) EN 300 175-4 as its standard.

DECT Forum (Digital Enhanced Cordless Telecommunications) Forum
An industry association that promotes DECT as a wireless communications technology. (http://www.dect.org/)

DECT ULE (Digital Enhanced Cordless Telecommunications Ultra Low Energy)
A DECT version that efficiently uses battery power for long life, and offers M2M application possibilities in wireless sensor networks for home, enterprise, and neighborhood networks.

DEED (Demonstration of Energy-Efficient Developments)
An APPA (American Public Power Association) R&D program for members that encourages activities that promote energy innovation, improve efficiencies, and lower costs of energy to public power customers.

DEER (Database for Energy Efficiency Resources)
A database that contains information on selected energy-efficient technologies and provides estimates of the energy-savings potential for these technologies compared to energy-intensive equipment. It was jointly developed by the CPUC (California Public Utilities Commission) and the CEC (California Energy Commission) with input from IOUs (Investor Owned Utilities) and other stakeholders. It is funded by California ratepayers under the auspices of the CPUC. (http://www.deeresources.com/)

Deferred income tax
A balance sheet liability showing additional federal income taxes that would have been due if a utility had not been allowed to compute tax expenses differently for income tax reporting purposes than for rate-making purposes.

Deionized water
Water that has inorganic chemicals removed through a distillation process, used for specific scientific processes. Also known as demineralized water.

Delivered energy
The amount of energy delivered to a site, also known as net energy.

Deliveries
Electricity generated by one system and delivered to another system through transmission lines.

Demand
The electric energy required at a given instant of time. It is typically calculated as an average requirement over a time interval ranging from minutes to an hour. It is also the measure of power that a load receives or requires. It is usually noted in kW.

Demand bid
A power exchange bid for a quantity of electricity that a customer or utility is willing to purchase and, if relevant, the maximum price to be paid.

Demand bid/buyback
A power exchange bid for a quantity of electricity that a customer, utility, or CSP (Curtailment Service Provider) is willing to curtail or reduce for a specific price. These transactions may be part of a demand response program and are most often found in wholesale markets. NERC (North

American Electric Reliability Corp.) classifies this as a form of dispatchable DR (Demand Response).

Demand charge
The portion of a customer's bill for electric service based on the maximum usage in that billing period and calculated based on the applicable rate schedule.

Demand charge credit
Compensation received by the buyer when the seller cannot meet the delivery terms of the contract.

Demand indicator
A measure of the number of energy-consuming units, or the amount of service or output, for which energy inputs are required.

Demand interval
The time period during which flow of electricity is measured.

Demand resources
Energy efficiency, load management, realtime demand response, and distributed-generation programs that will participate in the ISO NE (Independent System Operator New England) wholesale capacity market known as the FCM (Forward Capacity Market). The intent is to create a predictable revenue stream for capacity savings to enable better capital financing options and encourage other similar program rollouts.

Demand response programs
Utility programs designed to change on-site demand for energy through changes in prices, load control signals, or other incentives to customers. The programs are activated at times of peak usage or when system reliability is jeopardized. Demand response programs fall into two general categories known as price-based programs or capacity-based programs.

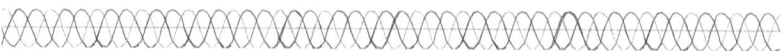

Price-based programs include dynamic pricing/tariffs, price-responsive demand bidding, and critical peak pricing structures that let users voluntarily reduce their electricity use. Capacity-based programs include contractually obligated reductions and direct load control/cycling. Utilities use these programs to address system reliability, asset use efficiency, and market conditions; and avoid investments in new T&D (Transmission and Distribution) assets, peaker plants, or expensive peak power purchases.

Demand response system

As defined in the ARRA Section 1306 amending EISA 2007 (Energy Independence and Security Act of 2007), software and hardware that automatically respond to price or other signals by adjusting electricity consumption in networked devices and then communicate results back to the signaling entity, such as a utility or service provider.

Demand-side management costs

The utility costs incurred from DSM (Demand-Side Management) activities. The costs are reported in the year in which they are incurred, not when any utility savings occur. Utility costs include all annual expenses (labor, administrative, equipment, incentives, marketing, monitoring, and evaluation), and the accounting treatment may be expensed or capitalized. DSM costs do not include lump sum capital costs or costs associated with strategic load growth.

Demonstration

The deployment of a new product or service for investigation of interaction with the utility's systems to ensure that the new item does not adversely impact utility operations from a reliability or quality perspective.

Deperimeterization

Security construct about the shift or elimination of typical physical and logical network boundaries due to the proliferation of mobile work, Web-based applications, and social media. It is addressed by using encryption and dynamic data-level authentication that essentially protects the data itself rather than the IT network or infrastructure.

DER (Distributed Energy Resources)
Grid-connected or standalone generation, energy storage, or negawatt assets that are deployed in the distribution grid. DER assets can substitute for or supplement grid-supplied power.

DERTF (Distributed Energy Resources Test Facility)
Part of NREL (National Renewable Energy Laboratory), this test facility is designed to assist the power industry in developing and testing distributed generation systems. Labs and outdoor test beds check the performance and reliability of these systems, support standards development, and investigate complex system integration issues. (http://www.nrel.gov/eis/test_facility_vtour.html)

DES (Data Encryption Standard)
A NIST (National Institute of Standards and Technology) standard data encryption algorithm. It has been superseded by AES.

Desalination
Processes that remove salts from water to create fresh water. Two of the most common processes are thermal distillation and reverse osmosis.

DESC (Digital Energy Solutions Campaign)
A nonprofit coalition of companies, associations, and environmental and energy non-governmental organizations advocating for increased energy efficiency through digital energy ICT (Information and Communications Technologies).

DESS (Distributed Energy Storage System)
An energy system design that contains both central and distributed energy storage technologies plus intelligent control to ensure the overall reliability of electricity delivery. It accommodates distributed generation through

peak shaving and power quality mitigation, enables islanding to reduce or eliminate brownouts and blackouts, and improves asset management plus voltage and frequency regulation. This system strategy can postpone or, ideally, eliminate the need for buildout of additional generating capacity.

DETL (Distributed Energy Technology Laboratory)
Part of SNLA (Sandia National Laboratory), this lab tests distributed generation technologies and control methodologies in the 15-kV class. This utility-interconnected test bed also provides hardware and controls for research aimed at improving SCADA (Supervisory Control And Data Acquisition) security, and is involved in microgrid research with distributed generation devices.

Dewatering
Any process that removes some percentage of water in sludge or slurry.

DEWG (Domain Expert Working Groups)
Teams working within NIST (National Institute of Standards and Technology) to provide technical input into the interoperability standards framework for NIST's intelligent electric grid work.

DFR (Digital Fault Recorder)
A device that captures transients and other disturbances to analyze substation events.

DG (Distributed Generation)
Electric generation that feeds into the distribution grid, rather than the bulk transmission grid, whether on the utility side or customer side of the meter. It does not require transmission assets. It includes customer-owned microturbines, wind-powered generators, hydro units, and PV arrays. Customers who own generation resources usually want to reduce the amount of power purchased from the local utility or supply their own backup power needs, and this form of DG is sometimes known as on-site

DG. Excess power may be sold back to the utility through net metering. Utilities may invest in DG to mitigate substation level peak loads and/or avoid building or upgrading local distribution lines. The technologies used in distributed generation are sometimes referred to as Distributed Energy Resources. Also known as decentralized energy.

DHS (Department of Homeland Security)
The federal agency responsible for security within the USA, including guarding against terrorism and improving readiness, response, and recovery from disasters. The DHS has taken a great deal of interest in cyber security and other electric grid security issues. (http://www.dhs.gov/index.shtm)

Diffie-Hellman
A cryptographic key exchange protocol named for two of its founders, Whitfield Diffie and Martin Hellman. The Diffie-Hellman key exchange protocol lets two parties who have no prior knowledge of each other establish a secret key over an insecure communication channel, which can then be used to encrypt and decrypt subsequent communications. Diffie-Hellman is considered one of the earliest known implementations of public-key cryptography.

DIGITALEUROPE
A European industry association comprised of IT, consumer electronics, and telecommunications sectors with a focus on regional public policy, legislation, and regulations. (http://www.eicta.org/)

Direct access
The ability of retail customers to purchase electricity directly from a supplier other than their traditional supplier.

Direct control load management
A utility's ability to interrupt power supply to individual appliances or equipment on customer premises. This type of control usually reduces the demand of residential customers during seasonal peak periods. It is demand-side management that is under the direct control of the utility.

Direct Load Control
See DLC.

Direct non-process end use
Use that is not specific to manufacturing operations. Non-process uses include HVAC, facility lighting, facility support, onsite transportation, conventional electricity generation, and any other operation that is not manufacturing-specific. It refers to electricity consumption.

Direct process end use
Use that is specific to manufacturing operations. Manufacturing-specific uses include process heating, process cooling and refrigeration, machine drive, and electrochemical processes. It refers to electricity consumption.

Direct use
Use of electricity that is self-generated and consumed by a service or industrial process located within the same facility or campus that houses the generating equipment.

Direct utility cost
A utility cost that is identified with a DSM (Demand-Side Management) program category.

Directive 95/46/EC
A European Union (EU) ruling also known as the European Data Protection Initiative, it protects the collection, use, and disclosure of personal information in private and public sectors with a particular emphasis on

online data. It requires organizations to provide easy access to each individual's own data, freedom to transfer data, rights to delete data, and informed consent to data processing. It has implications for Smart Grid energy and water consumption data as well as other M2M applications. It is national law in the EU member states.

Directive 2006/32/EC
A European Union (EU) ruling also known as the Energy Services or Energy Efficiency Directive, it creates a market for energy services and the delivery of energy efficiency programs to end users. It sets national energy savings targets for EU member governments and energy efficiency obligations on energy distributors or retailers. It is the motivation for smart meter deployments in Europe.

Discharge capacity
The amount of electricity that can be stored in an energy storage device.

Disinfection
A process used in surface water treatment plants to purify water by using various chemicals. Chlorine is the chemical most often used to disinfect water supplies.

Dispatchable demand response
Electricity consumption that can be reduced based on communication from a control center. It includes direct load control, interruptible demand, CPP (Critical Peak Pricing) with control, load as a capacity resource, spinning and non-spinning reserves, regulation, and energy-voluntary and energy-price resources.

Dispatchable power
Any combination of fossil fuel, nuclear, and renewable energies that maintains a predictable supply of power to a grid at the wholesale level.

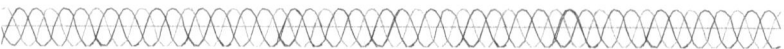

Dispatching
The operations of an integrated utility, including the assignment of load to specific generating stations, administration and maintenance of high-voltage lines, substations, and equipment, and scheduling of energy transactions with connecting electric utilities.

Distributed generation
See DG.

Distributed generation facility
A collection of equipment that generates and may store renewable electricity that is located near the customers that it serves. According to amendments to PURPA (Public Utilities Regulatory Policies Act), the maximum capacity of a facility is 2 MW.

Distributed generation management system
As defined in the ARRA Section 1306 amending EISA 2007 (Energy Independence and Security Act of 2007), a type of energy management system that responds to signals from a utility or service provider, controls the generating unit accordingly, and communicates results.

Distributed generator
An electricity generator located close to the particular load that it serves for the purpose of meeting substation peak loads and/or displacing the need to build or upgrade local distribution lines.

Distributed water heating system
A system for heating hot water located at more than one space within a building. A point-of-use water heater is located at the faucet and heats water for immediate use. These systems are used to improve energy efficiency. Also known as a point-of-use water heating system.

Distribution
The local delivery of electricity to customers.

Distribution automation
The range of applications and technologies, such as substation and feeder SCADA (Supervisory Control And Data Acquisition), OMS (Outage Management System) integration, and smart metering that add intelligence to improve reliability, efficiency, customer service, and asset management for utilities. It supports bi-directional energy and information flows.

Distribution facilities
A collection of components in a water supply system, these transport water from source to purification facility. The system components may be underground pipes, open aqueduct, or a covered tunnel system.

Distribution grid
The hardware and software systems that manage and deliver electricity from a substation to a customer, including power lines, transformers, wiring to meters, and meters. Distribution substations reduce power from high-voltage transmission lines to levels suitable for homes and businesses typically between 4kV and 55 kV. It is also called the distribution system.

Distribution provider
An entity that operates the electrical grid deployed between the transmission system and end users.

Disturbance
Any unplanned event or perturbation of the electric system that produces an abnormal system condition. Events can include a sudden generation failure or transmission problems.

Disturbance Control Standard
A defined time limit for a Balancing Authority to resolve a disturbance and return operations to within a specified range.

Diversity
The differences in loads between utilities based on time-of-day usage, facility usage, and/or demands placed upon the system grid.

Diversity exchange
A swap of capacity or energy between systems whose peak loads occur at different times.

Divestiture
The removal of a utility function by selling or changing the ownership of the assets related to that function. Typically, generation assets are divested so they are no longer owned by the shareholders that own the transmission and distribution assets.

DLC (Direct Load Control)
A capability that reduces electricity use by providing the system operator, utility, or CSP (Curtailment Service Provider) direct control to remotely shut down or cycle power to individual appliances or equipment on customer premises. It is a type of dispatchable demand response program that is under direct remote control of a control center, and is usually offered to residential (usually HVAC) and small commercial customers. It is generally used to balance supply and demand during peak periods or to avoid paying high prices for electricity during peak timeframes. It is sometimes referred to as direct electricity load control.

DLMS (Device Language Message Specification)
Formerly the Distribution Line Message Specification, it was established by IEC TC 57 and published as IEC 61334-4-41. It evolved to become the Device Language Message Specification with the objective to provide an

interoperable environment for structured modelling and meter data exchange.

DLMS User Association
An industry association that comprises manufacturers, users, and regulators focused on promoting the DLMS specification and creating generic and compatible communication objects for the integration of diverse metering systems. It maintains a conformance testing and certification scheme for testing implementations. (http://www.dlms.com/index2.php)

DLNA (Digital Living Network Alliance)
A global industry association of consumer electronics and computer and mobile device manufacturers that design wired and wireless products in an interoperable network by using open standards and widely available industry specifications that compose their Interoperability Guidelines. (http://www.dlna.org/home)

DMR (Digital Mobile Radio)
A digital radio standard for professional mobile radio users that was developed by ETSI (European Telecommunications Standards Institute). The standard operates within the existing 12.5 kHz channel spacing for licensed land mobile frequency bands globally and for future regulatory requirements for 6.25 kHz channel equivalence. It provides voice, data and other supplementary services. The protocol covers unlicensed (Tier I), licensed conventional (Tier II) and licensed trunked (Tier III) modes of operation.

DMRA (Digital Mobile Radio Association)
An industry association promoting the DMR standard through interoperability testing and certification programs; regulatory policy advocacy; and education. (http://dmrassociation.org/)

DMS (Distribution Management System)
See EMS.

DMTF (Distributed Management Task Force)
An organization of IT environment industry members that collaborate on the development, implantation, and promotion of IT infrastructure interoperability and management standards. (www.dmtf.org)

DNO (Distribution Network Operator)
A term for a company licensed to distribute electricity in the United Kingdom.

DNP Users Group
A nonprofit association promoting DNP3 protocol for use in the utility industry. It maintains documentation, facilitates technical discussion, and delivers education about DNP3. The Technical Committee specifies enhancements, ensures backwards compatibility and vendor compliance with the protocol. (http://www.dnp.org/)

DNP3 (Distributed Network Protocol 3)
A set of open, standards-based communication protocols used in SCADA (Supervisory Control And Data Acquisition) systems that enable different devices to exchange data and control messages. It is a robust communications protocol, but it is not secure. It is used in substations and between substations and operations centers. IEC 61850 may replace it in new substations.

DOC (Department of Commerce)
The federal agency with a broad mandate to advance economic growth and jobs for Americans. It oversees NIST (National Institute of Science and Technology) that has statutory authority to develop the Smart Grid standards road map and the International Trade Administration that works

to export USA technology, including Smart Grid solutions, globally. (http://www.commerce.gov/)

DOE (Department of Energy)
The federal agency responsible for advancing the national economic and energy security; promoting scientific and technological innovation in energy; and overseeing cleanup of the national nuclear weapons complex. It has unprecedented funding for the Smart Grid, energy storage, and energy sources, and uses its national laboratories, policies, and resources to promote energy solutions and develop policies and standards. Relevant agencies and labs are listed separately in this dictionary. (http://www.energy.gov/)

DOE Loan Guarantee Program
Section 1703 of Title XVII of the Energy Policy Act (EPAct) of 2005 created this loan program and was reauthorized and revised by the American Recovery and Reinvestment Act (ARRA) of 2009 with Section 1705. The 1705 Program was retired in September 2011, but the Section 1703 Program is active. Section 1703 authorizes the DOE to issue loan guarantees for projects that "avoid, reduce or sequester air pollutants or anthropogenic emissions of greenhouse gases; and employ new or significantly improved technologies as compared to commercial technologies in service in the United States at the time the guarantee is issued." The program is authorized to offer more than $10 billion in loan guarantees for energy efficiency, renewable energy and advanced transmission and distribution projects.

DOE Smart Grid Request for Information
A documented process to inform the DOE (Department of Energy) as it develops policies regarding the integration of broadband technologies with the Smart Grid. An initial RFI (Request for Information) received comments from electric utilities, consumer groups, and other stakeholders on public (federal and state) and private projects using consumer energy usage data

while protecting consumer privacy. A second RFI requested feedback on electric utilities' communications needs as Smart Grid technologies increasingly deploy across the country.

DoS (Denial of Service)
An Internet attack that attempts to disable a website by flooding it with requests so it cannot perform its regular functions. Also known as a distributed denial of service or DDoS.

DPF (Distribution Power Flow)
Algorithms and models that describe how electricity is distributed in a grid. DPF identifies the impacts of DER (Distributed Energy Resources) in the distribution grid for optimal grid management, including decisions based on market prices at the system level, and loss minimization at the distribution level.

DPWS (Devices Profile for Web Services)
Specifications that promote interoperability between Web services and clients, including messaging, discovery, description, and events on resource-constrained devices. It may be a key specification enabling M2M communications amongst a wide range of devices.

DR (Demand Response)
Behavioral changes in customer energy usage prompted by price changes or incentive payments to reduce consumption during events that range from addressing peak periods of demand, market conditions, or grid reliability conditions.

DR Ready (Demand Response Ready)
A capability built into appliances, building management systems, and other devices installed in homes and businesses to allow a device to receive and respond to signaling from a utility and modulate operation to reduce or shift demand.

DRAS (Demand Response Automation Server)
A server used in the OpenADR specification to facilitate the automation of customer response to DR programs and dynamic pricing through a communicating client.

Drawdown
Reduction of water level in surface or ground water supply based on pumping or other method of withdrawal. Also refers to a reduction in underground water pressure as a result of drilling activity.

DRE (Distributed Renewable Energy)
Also known as DRG (distributed renewable generation), this is the collection of clean fuel generation types deployed in low voltage distribution grids. Examples of these grid-connected generation sources are solar, wind, or tidal power, with solar being the predominant energy source today.

Drinking water
Water delivered to a consumer, after treatment that meets quality standards for human consumption.

Drinking Water State Revolving Fund
A federal fund in the USA that finances investments to improve compliance with drinking water standards. It was established by Congress in 1997.

DRIVE (Demand Response Impact and Value Estimation)
A model developed as part of the National Action Plan for Demand Response that estimates the impacts of demand response and Smart Grid programs on utility operations.

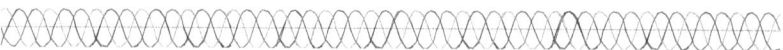

DRP (Demand Response Provider)
An entity that aggregates customers (residential, C&I, agricultural) and manages reductions or curtailments in electricity consumption or load as requested by utilities or ISO/RTOs.

DRPC (Dynamic Real Power Compensation)
A proposed technology that includes voltage source converters and large-scale fast response battery storage to deliver fast power compensation to accommodate power fluctuations in large operating ranges.

DRRC (Demand Response Research Center)
A research facility founded by the CEC (California Energy Commission) PIER (Public Interest Energy Research) group and hosted at Lawrence Berkeley National Laboratory. This research group issued an open communications standard and architecture for automated demand response systems in commercial and industrial facilities. The standard is designed to improve the reliability and cost-effectiveness of automating the responses of buildings to standardized electricity price signals. (http://DRRC.lbl.gov)

DRSG (Demand Response and Smart Grid Coalition)
A trade association of vendors that develop and deliver products and services for demand response, smart meters, and associated Smart Grid technologies. It provides education and information about how DR solutions work within the Smart Grid targeted to policymakers, utilities, the media, and investors. It was formerly known as the DRAM (Demand Response Advanced Metering Coalition). (http://www.drsgcoalition.org/)

Dry cooling system
Part of a thermoelectric generation technology, a system that uses either a direct or indirect air cooling process, using much less water than once-through or wet recirculating cooling processes.

Dry steam power plant
A type of geothermal power plant in which steam comes directly from a geothermal reservoir to turbines that power the generator. Because the reservoir only produces steam, no separation of water from steam is required. The other types of geothermal plants are binary power plant, flash power plant, and flash/binary combined cycle.

DSA (Digital Signature Algorithm)
An algorithm that creates a secure digital signature, but does not encrypt data.

DSA (Dynamic Spectrum Access)
A technology that locates unused or underutilized spectrum for dynamic sharing and manages assignments of spectrum to avoid interference. It has potential use in Cognitive Radio Networks to enable M2M ubiquity.

DSIRE (Database of State Incentives for Renewables & Efficiency)
A comprehensive source of information on federal, state, local, and utility incentives that promote renewable energy and energy efficiency. It is an ongoing project of the North Carolina Solar Center and the Interstate Renewable Energy Council funded by the DOE (Department of Energy). (http://www.dsireusa.org/)

DSM (Demand-Side Management)
Utility technology and program deployments designed to help customers change their electricity usage, including the timing of use and amount of use. These are activities that influence consumption behaviors and don't include changes based on government-mandated energy efficiency standards. DSM contains two components: energy efficiency and demand response.

DSPF (Distribution System Power Flow)
A mathematical model that assists in grid optimization planning and operations in distributed power systems. It will become even more important as the grid moves to a more distributed energy architecture. Calculations must account for a radial distribution grid, multiphase and unbalanced operation, unbalanced distributed load, multiple branches and nodes, and variable resistance and reactance values.

DTCR (Dynamic Thermal Circuit Rating)
Software that delivers real-time information about a transmission circuit's operating condition to assist in increasing and optimizing power flows.

DTT (Direct Transfer Trip)
The use of wired or wireless communications to transmit trip signals for equipment used in generation, transmission, or distribution of electricity. It plays an important role in ensuring worker safety and equipment protection in a grid that will include more distributed generation assets.

DWML (Digital Weather Markup Language)
A code language in an XML format that transmits National Digital Forecast Database (NDFD) data over the Internet.

Dynamic IP
A schema that changes device addressing every time that device logs into the network.

Dynamic Line Rating
The use of phasor data to monitor transmission line loadings and conduct realtime recalculations of line ratings along with weather data to derive actual ampacity of transmission lines.

Dynamic Pricing
Price structures that change based on when electricity is used. Some examples of dynamic pricing are real-time pricing, critical-peak pricing, and TOU pricing.

Dynamic Rating
The process of adjusting the thermal rating of power equipment based on factors such as air temperature, wind speed, and solar radiation to reflect actual operating conditions.

Dynamic Reserves
The amount of reserve that is available to preserve the system during frequency disturbance.

Dynamic Stability
Any situation that involves major disturbances, such as loss of generation, faults, and sudden load changes, over a longer time period than transient stability, typically several minutes. It is possible for control to affect dynamic stability even though transient stability is maintained. The action of turbine governors, excitation systems, and other devices can interact to stabilize or destabilize the power system several minutes after a disturbance has occurred.

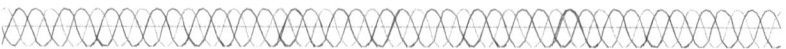

E

E-Tag
A NERC (North American Electric Reliability Corporation) term to identify an energy market transaction and its associated participants. An e-Tag can be thought of as a collection of transaction requests bundled together in one package and relating to a single transaction. A seven-character code is used as part of the e-Tag ID to identify a transaction.

E2SG
A European research project ending in 2015 that aims to reduce power losses in the grid by 20%.

EAC (Electric Advisory Committee)
A DOE (Department of Energy) advisory group focused on how to implement the Energy Policy Act of 2005, execute the Energy Independence and Security Act of 2007, and modernize the nation's electric grid. The most recent reports focus on the Smart Grid, energy storage technologies, and the adequacy of the electricity supply.
(http://www.oe.energy.gov/eac.htm)

EASE (European Association for Storage of Energy)
An industry association that promotes energy storage deployment and technological advancement in Europe and worldwide. Its primary mission is to implement a variety of energy storage technologies in the European energy system in order to improve grid flexibility and stability.
www.ease-storage.eu

Eastern Interconnection
One of the three AC power grids in North America covering a large section of Canada (east to central), the eastern USA, and across to the Rockies, excluding Texas. The Eastern Interconnection consists of the following

Reliability Councils or Regional Entities: Florida Reliability Coordinating Council, Midwest Reliability Organization, Northeast Power Coordinating Council, ReliabilityFirst Corporation, SERC Reliability Corporation, and the Southwest Power Pool. It is also a member of NERC (North American Electric Reliability Corporation). The following entities perform market operations in this region: PJM, NYISO, ISO-NE, IESO, MISO, and SPP.

ebIX (European forum for energy Business Information eXchange)
A European industry association with the mission to advance, develop, and standardize the use of electronic information exchange in the energy industry. The primary focus is exchanging internal energy market data for electricity and gas for both wholesale and retail markets. (http://www.ebix.org/)

ebXML (e business eXtensible Markup Language)
A modular suite of specifications to standardize XML business specifications. It enables a standard method to exchange business messages, conduct trading relationships, communicate data in common terms, and define and register business processes. It was approved as ISO 15000. It originated in 1999 with OASIS and the UNECE agency CEFACT.

ECC (Elliptic Curve Cryptography)
An approach to public-key cryptography based on mathematical structures of elliptic curves over finite fields. It is suitable for environments where computational resources are constrained in nature, and is used for AMI and ICS Smart Grid implementations, where efficiency is extremely important.

ECC (Electronic Communications Committee)
A CEPT (European Conference of Postal and Telecommunications administrations) committee that has influence in Smart Grid and M2M communications. It develops policies on European electronic communications activities, including cognitive radio, numbering, and other issues pertinent to M2M communications. (http://www.cept.org/ecc)

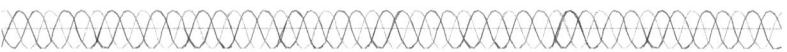

ECDSA (Elliptic Curve Digital Signal Algorithm)
An algorithm that creates a secure digital signature using public and private keys.

ECEDHA (Electrical and Computer Engineering Department Heads Association)
An organization of universities with accredited programs in electrical engineering and computer engineering focused on educational program advancement and development based on industry needs. It is managed by the International Engineering Consortium (IEC). (http://www.ecedha.org/)

Eclipse
An independent, not-for-profit organization originating from IBM whose mission is to encourage collaboration from individuals and organizations on building an open development platform that includes frameworks, tools, and runtimes for the full lifecycle of software. This software includes M2M applications for data reading and writing from sensors, actuators, large scale industrial processes, embedded systems, and mobile devices.

ECMA (European Computer Manufacturers Association)
An industry association focused on standardization of ICT (Information and Communications Technologies) and CE (Consumer Electronics) in cooperation with National, European and International standards organizations Standards. (http://www.ecma-international.org/)

EcoGridEU
A large-scale Smart Grid demonstration project that incorporates a real-time marketplace for DER (Distributed Energy Resources). It will operate with more than 50% of its energy coming from renewable sources, and it will encourage consumers to perform as prosumers. It is staged on the Danish island of Bornholm. (http://energinet.dk/en/forskning/EcoGrid-EU/sider/EU-EcoGrid-net.aspx)

Economic dispatch
The use of specific generation units to ensure the most economical production of electricity.

Economic energy
A regulatory term for excess energy from one utility that can be used by a neighboring utility in the same interconnection.

ECPA (Electronic Communications Protection Act)
A US federal law that was created to deliver balance between the privacy expectations of citizens and law enforcement needs. This law includes the Wiretap Act, the Stored Communications Act, and the Pen-Register Act. Smart Grid and M2M devices that communicate using IP addressing may be impacted by this legislation.

EDGE (Enhanced Data rates for GSM Evolution)
A high-speed mobile data standard based on GSM (Global System for Mobile Communication™) technology that delivers data at rates up to 384 Kbps using TDMA (Time Division Multiple Access). It is an enhancement for GPRS (General Packet Radio Service) networks with triple the data capacity. It can be deployed in many widely deployed bands, including 850 MHz, 900 MHz, 1800 MHz, and 1900 MHz.

EDSO-SG (European DSO Association for SmartGrids)
An industry association created by European distribution system operators.

EDTA (Electric Drive Transportation Association)
A USA industry association promoting electric drive for efficient and clean use of secure energy in the transportation sector through education, industry networking, public policy advocacy, and international conferences and exhibitions. Members include vehicle and equipment manufacturers, utilities, component suppliers, and end users. (http://www.electricdrive.org/)

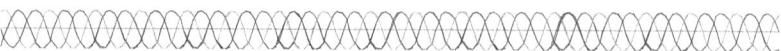

EE (Energy Efficiency)
The use of technology that requires less energy to perform the same function. It also encompasses programs that promote EE technology adoption and use. The EE objective is to reduce electricity consumption all the time through end-use energy solutions.

EEAC System (Energy Emergency Assurance Coordinator System)
A communications structure for state-level energy personnel to address energy emergencies or energy supply disruptions. It is part of the Office of Energy Assurance in the DOE (Department of Energy).

EECBG (Energy Efficiency and Conservation Block Grant)
A DOE (Department of Energy) program authorized in EISA 2007 (Energy Independence and Security Act of 2007) that assists USA cities, counties, states, territories, and Native American tribes in development, promotion, implementation, and management of energy efficiency and conservation projects and programs. It received ARRA funds of $3.2B. (http://www.eecbg.energy.gov/)

EEEL (Electronics and Electrical Engineering Laboratory)
Part of NIST (National Institute of Standards and Technology), it is tasked with creating metrology that allows new technologies to connect into existing power production and delivery systems and standards that protect the infrastructure from potential dangers introduced by deregulation. (www.nist.gov/eeel/quantum/applied_electrical/power.cfm)

EEGI (European Electricity Grid Initiative)
A European initiative within the SET-PLAN (Strategic Energy Technologies Plan) to conduct R&D to speed innovation for use in European electricity grids. Its goals include 35% of electricity from renewables by 2020 and completely decarbonized electricity production by 2050. (http://www.gridplus.eu/eegi)

EEI (Edison Electric Institute)
An association of USA-based IOUs (Investor Owned Utilities), representing almost 70 percent of the domestic electric power industry, and international electric utilities and industry manufacturers. It provides public policy leadership, industry information and intelligence, and educational opportunities. (www.eei.org)

EENS (Expected Energy Not Supplied)
A proposed performance metric for Smart Grid operations suggested by the NEMA (National Electrical Manufacturers Association).

EEPS (Energy Efficiency Portfolio Standards)
Programs to motivate and fund nonprofit organizations, public branches of government (such as K-12 public schools), and large private consumers to undertake efficiency upgrade projects that reduce net electricity demand.

EER (Energy Efficiency Ratio)
The instantaneous measurement of the cooling efficiency of an air conditioner or heat pump. The higher the EER, the more efficient the equipment.

EERA (European Energy Research Alliance)
An association of 10 leading European Research Institutes to accelerate the development of new energy technologies, particularly in renewables and energy storage, by conceiving and implementing research programs in support of the SET Plan (Strategic Energy Technology Plan). (http://www.eera-set.eu/home)

EERE (Energy Efficiency and Renewable Energy Office)
Part of the DOE (Department of Energy), this office manages programs to strengthen energy security, environmental quality, and economic vitality in public-private partnerships, including the NAPEE (National Action Plan for Energy Efficiency). (http://www.eere.energy.gov/)

EERS (Energy Efficiency Resource Standards)
A set of standards consisting of electric and/or gas energy-saving targets for utilities, often with flexibility to achieve the target through a market-based trading system. Through trading, a utility that saves more than its target can sell savings credits to utilities that fall short of their savings targets. Trading would also permit the market to find the lowest-cost savings. These standards vary from state to state. Currently, 19 states have enacted policies and laws to encourage more efficient use of electricity and natural gas based on EERS. ACEEE (American Council for an Energy-Efficient Economy) promotes EERS. (http://www.aceee.org/pubs/e063.htm)

EESAT (Electrical Energy Storage Applications & Technologies Conference)
A biennial, international conference that focuses on electrical energy storage technologies and power applications, including conventional and advanced battery energy storage, power electronics, SMES (Superconducting Magnetic Energy Storage), flywheels, CAES (Compressed Air Energy Storage), electrochemical capacitors, and pumped hydro. (http://www.sandia.gov/eesat)

EESCC (Energy Efficiency Standardization Coordination Collaborative)
An ANSI organization with the objective to assess the energy efficiency landscape; develop, and implement a standardization roadmap; and increase awareness and adoption of standards across public and private sectors. It consists of five working groups focused on building energy and water assessment standards; systems integration and communications; building energy modeling and labeling for energy performance; evaluation, measurement, and verification; and workforce credentialing. (http://www.ansi.org/standards_activities/standards_boards_panels/eescc /overview.aspx?menuid=3#.UMVPtqUmVUQ)

EETD (Environmental Energies Technology Division)
One of the primary research areas of Lawrence Berkeley National Lab focused on energy efficiency, storage, building efficiency, and policy. (http://eetd.lbl.gov/eetd.html)

EFET (European Federation of Energy Traders)
An industry association that promotes European energy trading in open, transparent, sustainable and liquid wholesale markets across national borders. (http://www.efet.org/)

Effluent
Wastewater from a sewage treatment or industrial plant.

EFRC (Energy Frontier Research Center)
Starting in 2009, the DOE's (Department of Energy) Office of Science is investing $777 million in 46 new research centers at universities, national laboratories, nonprofits, and private firms to accelerate scientific breakthroughs. Energy research will cover renewable and carbon-neutral energy, energy efficiency, energy storage, and cross-cutting science. Each center will receive $2 million to $5 million per year for an initial period of five years. (http://www.sc.doe.gov/bes/EFRC.html)

EGCC (Energy Government Coordinating Council)
A US government initiative composed of members from multiple agencies concerned with maintaining energy security through coordinated activities that include information sharing, identification of risks, and coordination and planning to mitigate risks.

EGHX (Exhaust Gas Heat eXchanger)
A system designed to remove heat from the exhaust gas of an engine and transfer it to a water circuit. This heated water can then be used for heating or electricity generation.

EGSA (Electrical Generating Systems Association)
A global trade association dedicated to onsite power generation. It functions as a source of information, education and training and promotes the development and monitoring of performance standards for the onsite power industry. (http://www.egsa.org/)

EHV (Extra High Voltage)
From a transmission perspective, usually more than 345 kV.

EHV (Electric hybrid vehicle)
An electric vehicle that consists of an electric system and an internal combustion system and is capable of operating on either system.

EIA (Energy Information Administration)
An office within the DOE (Department of Energy) that develops surveys, collects energy data, and analyzes and models energy issues. This agency provides information to Congress, other elements in DOE, FERC (Federal Energy Regulatory Commission), the Executive Branch, and industry groups and the public. It publishes EIA-861, the "Annual Electric Power Industry Report". (http://www.eia.doe.gov/)

EIEA (Energy Improvement and Extension Act)
Legislation enacted in 2008 that offered residential and commercial tax credits for renewable energy and energy efficiency purchases, extended production tax credits for wind, geothermal, landfill, and some biomass facilities, and a tax credit for plug-in EVs.

EII (European Industrial Initiatives)
European Commission-initiated joint large-scale technology development projects among academia, research, and industry. One of the six initiatives is focused on Smart Grids. The goal is to focus and align the efforts of the Community, member States, and industry to achieve common goals and to create a critical mass of activities and actors, thereby strengthening

industrial energy research and innovation on technologies for which working at the Community level adds most value. SETIS (Strategic Energy Technology Plan Information System) contributes to the definition of EIIs and Key Performance Indicators (KPIs) that are used to review and monitor EII progress.

EIPC (Eastern Interconnection Planning Collaborative)
A coalition of 24 transmission planning authorities in the Eastern USA and Canada that builds upon their individual regional expansion plans to develop a coordinated interregional analysis of transmission options and policy options. (http://eipconline.com/)

EIS Alliance (Energy Information Standards Alliance)
A consortium of equipment suppliers, energy consultants, and interested parties in the USA working together to develop common energy information standards, allowing homeowners, facility professionals, and manufacturing operations resources to intelligently manage energy consumption by enabling the sharing of energy information between equipment. The objective is to develop a common exchange framework for buildings to facilitate communication of common energy data. The EIS Alliance is involved with the federal government's Smart Grid Initiative and was been designated under PAP10 (Priority Action Plan 10) to develop use cases and requirement documents supporting the energy usage information in the customer domain. (http://www.eisalliance.org)

EISA 2007 (Energy Independence and Security Act of 2007)
Federal legislation passed in 2007 to codify a research, development, and demonstration program for Smart Grid technologies. Title XIII of EISA defined Smart Grid and authorized a number of programs, including NIST (National Institute of Standards and Technology) standards process. The 2009 Stimulus Bill amended Section 1304b to allow the DOE (Department of Energy)to fund up to 50 percent of Smart Grid demonstration project costs without any cap (previously the cap was $500 million). The 2009

Stimulus Bill also amended Section 1306 to now fund up to 50 percent of qualifying Smart Grid investments.

EISPC (Eastern Interconnection States' Planning Council)
An organization formed from state utility regulators and Governors' offices within the Eastern Interconnection to study transmission development within its boundaries. It includes 39 states and planning authorities such as PJM, NYISO, ISO-NE, SPP, and MISO.

ELC (Electrification Leadership Council)
An industry coalition of key stakeholders throughout the EV (electric vehicle) supply chain focused on working with federal, state, regional and local agencies and coalitions, utility companies, vehicle manufacturers and financial institutions to enable adoption and deployment of EVs and charging infrastructure.(http://www.elcouncil.com/)

ELCON (Electric Consumers Resource Council)
A national industry association representing large industrial users of electricity on federal and state policies that affect the price, availability, and reliability of electricity service. ELCON represents industrial electricity users with FERC (Federal Energy Regulatory Commission) and NERC (North American Electric Reliability Corp.). (http://www.elcon.org/)

Electric or electrical energy
The quantity of electricity delivered over a period of time. The commonly used unit of electrical energy is the kilowatt hour (kWh).

Electric generation industry
Fixed and portable generation industry that includes utilities, independent power producers, industrial and commercial power generators, and CHP (Combined Heat and Power) producers, but excludes single-family dwelling units.

Electric industry restructuring
The reconfiguration of vertically integrated electric utilities into markets with competing sellers, allowing customers to choose their suppliers but still receive delivery over the power lines of the local utility. Generation is now generally competitive, transmission is regulated by FERC (Federal Energy Regulatory Commission), and distribution falls under state jurisdictions.

Electric meter
The utility device that measures and records electricity use that is retrieved by manual means (including drive-by) to calculate customer bills. The devices may be solid state or electromechanical. It is sometimes referred to as the utility's cash register.

Electric operating expenses
The total of electric operation-related expenses, such as operation expenses, maintenance expenses, depreciation expenses, amortization, taxes other than income taxes, federal income taxes, other income taxes, provision for deferred income taxes, provision for deferred income-credit, and investment tax credit adjustment.

Electric or electrical power
The rate of delivery of electrical energy and the most frequently used measure of capacity. The basic unit is the kW (kilowatt) and generation and transmission are usually expressed in MW (Megawatt) or GW (Gigawatt).

Electric power grid
A system of synchronized power providers and consumers connected by transmission and distribution lines and operated by one or more control centers. The continental USA electric power grid consists of three systems: the Eastern Interconnect, Western Interconnect, and Texas Interconnect.

Electric power plant
A station containing prime movers, electric generators, and auxiliary equipment for converting mechanical, chemical, and/or fission energy into electric energy.

Electric power sector
An industry sector that consists of electricity-only and CHP (Combined Heat and Power) plants whose primary business is to sell electricity, or electricity and heat, to customers. The electricity can be produced by any form of energy.

Electric power system
Another term for the electrical grid.

Electric rate
The price set for a specified amount and type of electricity by class of service in an electric rate schedule.

Electric rate schedule
The documented electric rate and all terms and conditions approved by the applicable regulatory authority for all utility customers.

Electric Rule 21
This is a regulation developed by the California Public Utilities Commission (CPUC) that addresses the interconnection, operation, and metering requirements that electricity generating facilities must meet to be connected to a utility's electricity distribution grid.

Electric system loss
The total energy loss from all causes for an electric utility.

Electric system reliability
Adequacy and security of electricity. Adequacy means there are generation and transmission resources available to meet projected electrical demand

plus reserves for contingencies. Security means the system has sufficient operating capacity even after outages or other equipment failure.

Electric transportation
Any personal or mass transit vehicle propelled by a stationary or onboard electricity source.

Electric transportation technology
As defined in the ARRA Section 1306 amending EISA 2007 (Energy Independence and Security Act of 2007), any electric motor technology that provides some or all the motive power for a vehicle, including BEVs (Battery Electric Vehicles), EVs (Electric Vehicles), PHEVs (Plugin Hybrid Electric Vehicles), and fuel cell EVs, for personal or mass transit.

Electric utility
An entity that generates, transmits, or distributes electricity and recovers the cost of its generation, transmission or distribution assets, and operations through cost-based rates set by a separate regulatory authority. Utilities may be investor-owned, owned by a governmental unit (federal, municipal, or district), or owned by the consumers that the entity serves. Also known as an electric power system.

Electric utility divestiture
The separation of one electric utility function from others through the selling of the management and ownership of the assets related to that function. It generally means selling generation assets from transmission and distribution assets. It is one outcome of electric industry restructuring.

Electrical storage management system
As defined in the ARRA Section 1306 amending EISA 2007 (Energy Independence and Security Act of 2007), a software and hardware system that provides realtime and remote storage and discharge of electricity or thermal energy and reacts to signals to adjust battery operations and

communicates status and results back to a utility, service provider, or microgrid owner.

Electrical transmission system
The part of the electric grid that transports electricity at high voltages from generation facilities to substations. It includes the transmission lines and associated equipment, interconnections to neighboring systems, software applications used to manage and monitor the movement of electricity, and all sensing equipment in the transmission grid.

Electricity
A form of energy characterized by the presence and motion of elementary charged particles generated by friction, induction, or chemical change.

Electricity Directive 2003/54/EC
The key European legislation creating the internal market of electricity. It establishes common rules for the generation, transmission and distribution of electricity. It also defines rules for the organization and functioning of the electricity sector, access to the market, the criteria and procedures applicable to calls for tenders and the granting of authorizations and the operation of systems.
(http://eurlex.europa.eu/LexUriServ/LexUriServ.do?uri=OJ:L:2003:176:003 7:0055:EN:PDF)

Electricity sales
The number of kilowatthours sold in a given period of time to a class of service, such as residential, commercial, industrial, and other classes (sales to public authorities and sales to transportation).

Electricity supply chain
A conceptual construction of the electrical grid that applies from microgrids up to regional and national grids starting at electricity production and ending at customer use.

Electrification Coalition
A not-for-profit group of business leaders that promotes policies and actions for mass-scale deployment of electric vehicles for reasons of economic, environmental, and national security – all threatened by the dependence on oil. (http://www.electrificationcoalition.org/)

Electronic Frontier Foundation
A nonprofit organization focused on representing public interests in digital rights situations, which includes data privacy and the impacts that Smart Grid and M2M communications have on it. (https://www.eff.org/)

EM&V (Evaluation, Measurement & Verification)
A process to measure the performance and benefits of energy efficiency programs. It includes assessment of a program rollout, identification of costs and savings through standard metrics, and estimations of future impacts based on changes in energy market conditions. No standard EM&V process exists today, but several organizations have recommended procedures that are widely adopted for use.

Embedded security
Security implemented at the lowest level possible – such as the microchip level.

Embedded wireless
A term that describes the factory installation of wireless communications technologies in devices that are not phones.

Embodied Water
The assessment and calculation of the total water used in the production, transportation and consumption of an item or service. For example, it takes almost 265 gallons (1000 liters) of fresh water to produce just over one quart (1 liter) of milk. It also includes total water polluted over the item or service lifecycle. Also known as embedded water or virtual water.

EMCS (Energy Management and Control System)
An energy conservation system that uses computers, instrumentation, control equipment, and software to manage a building's use of energy for HVAC, lighting, and/or business-related processes. These systems can also manage fire control, safety, and security, but do not include time-clock thermostats. EMCS solutions are most often found in C&I (Commercial and Industrial) applications.

EMerge Alliance
An industry association of building owners, product manufacturers, and architectural firms that promotes the rapid adoption of safe, low-voltage DC power distribution and use in commercial building interiors. EMerge is focused on developing an open standard that integrates interior infrastructures, power, controls and a wide variety of peripheral devices, such as lighting, in a common platform.
(http://www.emergealliance.org/en/index.asp)

Emergency
The failure of an electric power system to generate or deliver electric power as normally intended, resulting in the cutoff or curtailment of service.

Emergency backup generation
The use of electric generators only during interruptions of normal power supply.

Emergency demand response
A demand response program that offers payments to customers for reducing their loads during reliability-triggered events. There is no payment or penalty if the customer does not curtail. It is most often found in wholesale markets.

Emergency energy
Electric energy provided for a limited duration, intended only for use during emergency conditions.

EMF (Electromagnetic Fields)
Electromagnetic fields are created by electric power charges and are present whenever and wherever electricity is generated, transmitted, distributed, or consumed. There are two types of fields: electric fields, which result from the voltage of the charge, and magnetic fields, which result from the amperage of the charge.

EMI (Electromagnetic Interference)
Any disturbance that impacts an electrical circuit because of electromagnetic emissions from an external source. Also known as RFI (Radio Frequency Interference).

EMIX (Energy Market Information eXchange)
A specification developed by an OASIS Technical Committee that defines an information model for power and energy markets to exchange price and product information. It includes quantity and quality of supply and distinguishes between power and energy sources.

EMP (Electromagnetic Pulse)
A natural geomagnetic storm (solar flare/wind) or man-made (nuclear detonation) phenomenon that stresses electrical components and systems, potentially damaging or destroying telecommunications and power delivery grids.

EMRI (Electricity Market Reform Initiative)
An APPA (American Public Power Association) initiative focused on investigation of the restructured wholesale electricity markets and development of proposals for identified reforms to the markets.

EMS (Energy Management System)
An application software suite that supports electric system operations, providing information and tools to operations and engineering staff to ensure reliable delivery of energy to customers. It is also known as a DMS

(Distribution Management System) for the bulk power system and sometimes refers to a residential software system that coordinates the operation of smart residential appliances, which can control loads (smart thermostats, water heater controls, smart refrigerators, and laundry equipment), along with on-site power generation, such as solar photovoltaics) and electrical energy storage. In residential situations, it is commonly known as HEMS (Home Energy Management System).

Encryption
A cryptographic process that makes text information unreadable to anyone or anything that lacks the decoding algorithm or "key."

Energy
The capacity for doing work measured as potential energy or kinetic energy. Electrical energy is usually measured in kilowatthours, or kWh.

Energy 2020
The European Union energy strategy that outlines the following five priorities: making Europe energy-efficient; building a pan-European integrated energy market; empowering consumers; extending their leadership in energy technology and innovations; leveraging the strong EU energy market to influence world markets. (http://ec.europa.eu/energy/publications/doc/2011_energy2020_en.pdf)

Energy arbitrage
The business of using stored energy purchased during low rates and selling it during higher-rate periods.

Energy assurance
Federal initiative that coordinates with state and local government levels plus private industry to strengthen and expand plans for resiliency of energy grids and supplies, improved energy security, and enhanced emergency response through integration of new energy sources, Smart

Grid applications, and coordination and communication capabilities.
(http://energy.gov/oe/services/energy-assurance)

Energy audit
A program carried out by a utility company or an affiliate in which an
auditor inspects a home and suggests ways energy can be saved.

Energy Access Practitioner Network
Part of the UN's Sustainable Energy for All initiative this program focuses on
household and community-level electrification.
(http://www.sustainableenergyforall.org/events-outreach/practitioner-
network)

Energy charge
The part of an electric bill based on the electric energy consumed or billed.

Energy conservation
Behavior that results in a reduction in energy use. Energy conservation is
different from demand response, which is based on deliberate adjustments
to usage to achieve reduced electricity demand. It is also different from
energy efficiency, which is optimization of technologies to perform the
same with less energy consumption.

Energy density
The energy stored per unit volume, identified as Wh/L, used to describe
attributes of energy storage technologies.

Energy efficiency
Technologies, applications, and services that reduce the consumption of
energy without impacting operations or behaviors.

Energy efficient motor
A motor virtually interchangeable with a standard motor, but a difference in construction makes it more energy efficient.

Energy harvesting
This is the process of capturing ambient energy in the local environment, such as heat, light, sound, vibration, or movement. Devices that harvest or scavenge energy can capture, accumulate, store, condition, and manage this energy into electricity for later use in performing a small, low energy task. The most common application of energy harvesting is powering wireless sensor networks, which are otherwise limited by battery life or lack of access to a plug to a power grid.

Energy intensity
The ratio of consumption to floor space.

Energy Interoperation (EI) V1.0
This is an information model that relies on the OASIS EMIX specification that enables energy transactions via a standardized exchange of market and generation signals. The specifications in the model are built in an XML language that allows signal interoperability as well as collaborative and transactive use of energy.

Energy – price resources
Customers who allow their energy providers to cut off their power supply during shortages in return for a payment that reflects the price of electricity they did not use. These resources are classified by NERC (North American Electric Reliability Corporation) as a form of dispatchable DR (Demand Response).

Energy shifting
Deliberate actions to use appliances and electronics that consume large electricity loads at times when rates are lowest or away from peak demand periods.

Energy Star®
A voluntary testing program sponsored by the EPA (Environmental Protection Agency) and DOE (Department of Energy) to help businesses and individuals protect the environment through use of labeled products with improved energy efficiency. The EPA is responsible for appliances, and the DOE is responsible for building standards. The label is available on over 60 product categories, including major appliances, office equipment, lighting, and home electronics. It is also an option for new homes and commercial and industrial buildings. (http://www.energystar.gov/)

Energy storage
Banking electricity in batteries or other storage devices for transmission and distribution in the future. Storage can take electrochemical or non-electrochemical forms. Energy storage is used for grid stabilization, grid operational support (frequency regulation services, contingency reserves, voltage support, and black start), power quality and reliability, load shifting, and compensating for the intermittent nature of renewable energy sources like wind and solar. Energy storage is a key component of the Smart Grid.

Energy surety
The ability to continue operations in response to natural and man-made disasters by ensuring the reliability, security, and resiliency of sources of energy and corresponding infrastructure at the national, state, and local levels.

Energy sustainability
As defined in the ARRA Section 1306 amending EISA 2007 (Energy Independence and Security Act of 2007), a concept that drives innovative technologies and processes for production of electricity and transportation fuels that are environmentally benign, reduce greenhouse gas production, and use renewable or bio-based resources that are abundant in the world.

Enhanced geothermal systems
Any underground reservoir system that was engineered instead of occurring naturally. It is designed to artificially create permeability in hot rock and introduces water or another working fluid to extract the heat and produce electricity or direct heat.

ENISA (Energy Efficiency Standardization Coordination Collaborative)
A European Union governmental agency with a mission to develop a culture of network and information security through the exchange of information, best practices and knowledge in information security. (http://www.enisa.europa.eu)

EnR (European Energy Network)
A voluntary network of 23 European energy agencies with responsibility for the planning, management or review of national research, development, demonstration or dissemination programs in the fields of energy efficiency and renewable energy and climate change abatement. EnR shares information through eight Working Groups, which are also open to relevant non-member organizations. The Working Groups are Behavior Change, Energy Efficiency, Buildings, Transport, Renewable Energy, Monitoring Tools, Labeling and Ecodesign, and Financing Sustainable Energy. (http://www.enr-network.org/)

ENTSO-E (European Network of Transmission System Operators for Electricity)
A European industry association that replaced the six Transmission System Operator associations and the Union for the Coordination of Transmission of Electricity (UCTE), responsible for region-wide reliable operation, optimal management, and sound technical evolution of the European electricity transmission system. (http://www.entsoe.eu/)

Environment Canada
The federal department in Canada responsible to preserve and enhance the quality of the natural environment, including water, air, soil, flora and fauna and conserve and protect Canada's water resources, along with other objectives. (http://www.ec.gc.ca/)

EPA (Energy Policy Alliance)
A coalition of engineering societies focused on the links between energy and national security and promotion of the commercialization of new energy technologies and basic scientific and energy research in the USA (http://www.energypolicyalliance.org/home.cfm)

EPA (Environmental Protection Agency)
The USA federal agency that leads national environmental science, research, education, and assessment efforts. The EPA protects human health and the environment and it has regulatory authority and enforcement over a number of toxins, including CO_2 emissions. EPA contains an Office of Water that handles the agency's water quality activities, protecting, regulating and improving the country's water supplies. (http://www.epa.gov/)

EPA Tier 3 Standards
Emission standards set by the EPA (Environmental Protection Agency) for diesel generator sets running in emergency situations. The standards also include requirements for diesel fuel, but are relaxed from Tier 4 standards.

EPA Tier 4 Standards
Emission standards set by the EPA (Environmental Protection Agency) for diesel generator sets running in non-emergency situations. The standards also include requirements for diesel fuel to decrease sulfur levels by more than 99%.

EPAct 1992 (Energy Policy Act of 1992)
Federal legislation that created a new class of power generators known as exempt wholesale generators, exempt from the provisions of the Public Holding Company Act of 1935 and granted authority to the FERC (Federal Energy Regulatory Commission) to order and condition access by eligible parties to the interconnected transmission grid. It also established water efficiency standards for appliances, including reducing the amount of water used by new toilets to 1.6 gallons/flush.

EPAct 2005 (Energy Policy Act of 2005)
Federal legislation passed in 2005 that, in Section 103, mandated a 20 percent energy efficiency improvement and advanced metering for federal buildings. In Section 1252, it also requires utilities to offer net metering to customers upon their request and outlines that metering must be time-based, capable of supporting TOU (Time Of Use), CPP (Critical Peak Pricing), and hourly real-time pricing structures, and have communications, or explain non-compliance. This act also granted NERC (North American Electric Reliability Corp.) authority to impose fines for non-compliance with NERC's reliability standards, subject to FERC (Federal Energy Regulatory Commission) oversight.

EPCE (Energy Providers Coalition for Education)
A US-based alliance focused on online energy education. (http://www.epceonline.org/)

EPIC (Electric Program Investment Charge)
A fund created by the CPUC (California Public Utilities Commission) to provide public interest investments in applied research and development; technology demonstration and deployment; market support, and market facilitation of clean energy technologies and approaches for the benefit of electricity ratepayers of the state's three IOUs (investor-owned utilities). The utilities are Pacific Gas and Electric Company (PG&E), San Diego Gas & Electric Company (SDG&E), and Southern California Edison (SCE). 80% of the funding is administered by the CEC (California Energy Commission) and 20% by the utilities.

EPIC (Electronic Privacy Information Center)
A USA-based nonprofit organization established to focus public attention on emerging civil liberties issues and to protect privacy, the First Amendment, and constitutional values. (http://epic.org/)

EPoSS (European technology Platform on Smart Systems integration)
A European industry-driven policy initiative of public and private stakeholders focused on smart systems integration and integrated micro- and nanosystems. Objectives include formation of a roadmap, a strategic research agenda, and definition of research priorities. (http://www.smart-systems-integration.org/public)

EPRI (Electric Power Research Institute, Inc.)
A nonprofit organization funded by global energy companies, utilities, and technology providers, that conducts research and development relating to the generation, delivery, and use of electricity for the benefit of the public. Focused on addressing challenges in electricity, including reliability, efficiency, health, safety, and the environment, it also provides technology, policy, and economic analyses for long-range research and development planning and supports research in emerging technologies. EPRI is working with NIST (National Institute of Standards and Technology) to build an initial set of Smart Grid standards. Members represent more than 90 percent of the electricity generated and delivered in the United States, and international participation extends to 40 countries. EPRI's principal offices and laboratories are located in Palo Alto, CA Charlotte, NC; Knoxville, TN; and Lenox, MA. (www.epri.com)

EPSA (Electric Power Supply Association)
A national industry association of power generators and marketers that advocates for policies and practices focused on competitive market guidelines and for the formation of RTOs (Regional Transmission Organizations) to expand competitive power transactions in those US regions that do not have them. EPSA's members own more than 480

electricity generation plants representing 200,000 MW of capacity in 40 states and the District of Columbia. (www.epsa.org)

EPSMA (European Power Supply Manufacturers Association)
An industry association focused on the needs of the switched-mode and linear power supply industry in Europe.
(http://www.epsma.org/default.htm)

ERCOT (Electric Reliability Council of Texas)
The ISO (Independent System Operator) that manages the deregulated market for 75 percent of the state of Texas. ERCOT schedules and dispatches power on an electric grid that connects 38,000 miles of transmission lines and more than 550 generation units. ERCOT also manages financial settlement for the competitive wholesale bulk-power market and administers customer switching in competitive choice areas. It is one of the 10 ISO/RTOs (Regional Transmission Organizations) currently operating in North America. (www.ercot.com/)

EREV (Extended Range Electric Vehicle)
Also known as a REEV (Range-Extended Electric Vehicle), this is an all-electric vehicle that includes a small generator, which could be gas-powered, to recharge the vehicle batteries on longer trips.

ERGEG (European Regulators' Group for Electricity and Gas)
This organization advises and assists the European Commission through draft measure recommendations that facilitate the creation of a single, competitive, efficient and sustainable internal market for gas and electricity in Europe. Their Regional Initiatives establish seven electricity and three regional gas markets in Europe as an intermediate step to the creation of a single, competitive EU market in electricity and gas.
(http://www.energyregulators.eu/portal/page/portal/EER_HOME/EER_AB OUT/ERGEG)

ERO (Electricity Reliability Organization)
The name used in the US Energy Policy Act of 2005 to refer to the independent entity that is given the authority to develop and enforce mandatory reliability standards for the North American bulk power system. NERC (North American Electric Reliability Corporation) was designated as the electricity reliability organization by FERC (Federal Energy Regulatory Commission) in 2006. As a result, a significant number of NERC's existing 115 electric reliability standards will become mandatory and enforceable under the Federal Power Act. Another outcome of this designation is that utilities are now required to submit information and data to NERC. (http://www.nerc.com/)

ERRA (Energy Regulators Regional Association)
An organization comprised of independent energy regulatory bodies primarily from the Central European and Eurasian region, with affiliates from Asia, the Middle East, and the USA. The main objectives are to exchange of information and experience among members, improve national energy regulation in member countries, and foster development of energy regulators. (http://www.erranet.org/)

ERTICO
A public/private partnership with members from the European Commission, ministries of transport, and European businesses pursuing the development and deployment of intelligent transport systems and services. It advances telematics services and geo-localization services, as well as safety and environmental initiatives. (http://www.ertico.com/)

ESIC (Energy Storage Integration Council)
An EPRI-sponsored forum for electric utilities to develop safe, cost effective energy solutions for utility use in collaboration with energy storage vendors, policy-makers, and other stakeholders. There are four working groups focused on applications, performance, system development, and grid deployment.

ES-ISAC (Electricity Sector Information Sharing and Analysis Center)
A NERC (North American Electric Reliability Corporation)-operated group under the US Department of Homeland Security and Public Safety Canada. ES-ISAC gathers information about security-related threats and incidents in the electricity sector and communicates it to electricity sector, government, and other critical infrastructure participants. Threats can be physical or cyber. (http://www.esisac.com/)

ES3PJWG (Energy Sector Public-Private Partnership Joint Working Group)
A public-private partnership of governmental agency representatives, natural gas and electric utility resources and cyber-security professionals. It provides a platform to build awareness encourage innovations that improve the cyber security of energy infrastructure control systems.

ESA (Electricity Storage Association)
A trade association established to foster development and commercialization of energy storage technologies. It promotes the development and commercialization of competitive and reliable energy storage delivery systems. It is focused on bulk power as well as distributed energy storage, and members represent utilities, independent power producers, energy storage manufacturers, and researchers. It was formerly called the Energy Storage Association. (http://www.electricitystorage.org/)

ESC (Energy Solutions Center)
A technology commercialization and market development organization that comprises energy utilities, public gas systems, and equipment manufacturers and vendors focused on gas-fueled technologies. (www.energysolutionscenter.org)

ESCO (Energy Service Company)
Any entity that develops, installs, and arranges financing for facility energy efficiency and maintenance optimization projects. They are project

developers that take the technical and performance risk associated with their energy efficiency projects.

ESDM (Energy Surety Design Methodology)
A quantitative risk-based assessment tool enabling communities to evaluate energy needs, improve the reliability and resiliency of their electric grids, and develop economical strategies to achieve energy surety. ESDM incorporates smart grid technologies and DER such as backup generators, solar power, and energy storage. Developed by Sandia National Lab, ESDM is proven to improve grid reliability and resiliency with renewables and DER.

ESI (Energy Services Interface)
A NIST (National Institute of Standards and Technology) term for the information management gateway for interactions between a consumer and energy service providers, such as electric utilities, or any entities that aggregate demand response, energy management services, and other such offerings. ESI functions include demand response signaling customer energy usage information to HEMS (Home Energy Management Systems) and IHDs (In Home Displays).

ESMA (European Smart Metering Alliance)
A project formed by several entities with partial funding from the European Union's Intelligent Energy program to research smart metering activity. This alliance is now closed because the project was concluded.

ESMIG (European Smart Metering Industry Group)
European industry consortium of energy companies to share smart metering information and expertise and promote the transformation of the European power grid into a Smart Grid complete with the latest technologies. (http://www.esmig.eu)

ESNA (Energy Services Network Association)
A not-for-profit industry association promoting the application of advanced energy management systems, including AMR/AMM (Automatic Meter Management), based on the NES (Network Energy Services) platform and the interoperability standard for utility networks NTA 8150. Members are primarily based in Europe. (http://www.esna.org)

ESP (Electronic Security Perimeter)
An important component of the CIP (Critical Infrastructure Protection) standards to secure critical cyber assets in the electrical grid. Standard CIP-005 requires the identification and protection of the Electronic Security Perimeter(s), inside which all Critical Cyber Assets reside, and all access points on the perimeter.

ETAC (Energy Technology Assessment Center)
An EPRI (Electric Power Research Institute) entity that conducts strategic assessments of electricity sector technology needs. ETAC research focuses on interdisciplinary analysis of technology development energy policy and economic factors. (http://my.epri.com/portal/server.pt?open=514&mode=2&objID=410)

ETCC (Emerging Technologies Coordinating Council)
An association of CEC (California Energy Commission) and California IOUs (Investor Owned Utilities) to promote cost-effective, energy-efficient technologies. The ETCC assists progression from R&D to product rollout for promising technologies. It provides an information exchange for the five stakeholder organizations about opportunities and results from Emerging Technologies activities. The CPUC (California Public Utilities Commission) finances ETCC operations from IOU ratepayer Public Goods Charge funds. (www.etcc-ca.com.)

Ethernet
Ethernet is the common name for IEEE 802.3, a packet-based transmission protocol that works on wired and wireless networks. Wired speeds range from 1Mbps – 1Gbps, and wireless speeds range from 2Mbps – 11Mbps.

ETIS (global IT association for telecommunications)
An association of major European telecommunications providers focused on improving their business performance by personal exchange of information on using ICT effectively. It includes 9 working groups dedicated to topics including information security, billing, enterprise architecture, and business intelligence. (http://www.etis.org/)

ETRM (Energy Trading and Risk Management)
Software applications, architecture, and tools that enable and track energy trading activity. The software helps manage the front, middle, and back office aspects of an energy trading entity.

ETSI (European Telecommunications Standards Institute)
A not-for-profit organization that produces globally applicable standards for information and communications technologies, including fixed, mobile, radio, converged, broadcast, M2M, and Internet technologies. It has almost 700 member organizations from 60 countries worldwide. It produced the Hiperman standard (High Performance Radio Metropolitan Area Network, a European alternative to WiMAX.) (http://www.etsi.org/)

ETSI OSGP (European Telecommunications Standards Institute Open Smart Grid Protocol)
Specifications created by ESNA (Energy Services Network Association) and published by ETSI to promote open applications based on the NES (Network Energy Services) architecture. OSGP supports the communication requirements between devices and a utility supplier or suppliers for data collection.

Eurelectric
A European electricity industry association formed as a merger of UNIPEDE and EURELECTRIC in 1999. Its mission is to contribute to the development and competitiveness of the electricity industry and to promote the role of electricity in the advancement of society. It represents the electricity industry in public affairs within the EU. (http://www.eurelectric.org/)

EUROGI
A public/private partnership organization in Europe with a mission to maximize the availability and effective use of geographic information throughout that region. It advocates for relevant technologies and policies. (http://www.eurogi.org/)

EV (Electric Vehicle)
Cars or other vehicles that draw electricity from batteries to power an electric motor. The batteries in an EV are charged using standard household electricity and electricity captured by regenerative braking. From a utility's perspective, an EV is equivalent to one house circuit.

EV-DO (Evolution Data Only *or* Optimized)
A 3G broadband wireless technology operating at 300 to 600 kbps with a bandwidth of 1.25 MHz. Its official name is CDMA2000 - High Rate Packet Data Air Interface. It is one of two primary 3G standards. The competing standard is W-CDMA. Verizon Wireless and Sprint deploy EV-DO in their networks.

EVO (Efficiency Valuation Organization)
A global, nonprofit organization dedicated to creating measurement and verification tools for efficiency programs. It is responsible for development of the IPMVP overview of best practices. (http://www.evo-world.org/)

EVSE (Electric Vehicle Supply Equipment)
The ungrounded, grounded, and equipment grounding conductors, EV connectors, plugs, and other apparatus deployed to deliver charge from a home or building. EVSE does not include battery-charging components such as rectifiers or transformers.

EWG (Exempt Wholesale Generator)
The 1992 Energy Policy Act created Wholesale Generator entities that are exempt from specific financial and legal restrictions of the Public Utilities Holding Company Act of 1935. These are independent power producers.

Ex Ante
A regulatory term that describes energy efficiency program savings as one of three levels. *Ex ante* means savings estimated or projected by the program design. The other two measures are gross savings and net savings.

Exceptional dispatch
A CAISO term for a manual override of software models to dispatch a generating unit in response to conditions that impact grid system reliability and ancillary services testing, voltage support, overgeneration mitigation, and other situations that require immediate action.

Export metering
The ability of a smart meter to measure electricity that is backfeeding the utility to support price structures for distributed generation and microgrids.

Extraordinary income deductions
Items of a non-recurring nature that are not typical or customary utility business activities, which would significantly modify the current year's net income if not treated as extraordinary items.

F

Facility
An existing or planned location that houses equipment that converts energy into electric energy.

Facility Smart Grid Information Model (BSR/ASHRAE/NEMA SPC201P)
This data model standard sponsored by ASHRAE and NEMA enables energy consuming devices and control systems in the customer premises to manage electrical loads and generation sources (primarily from commercial and industrial facilities) in response to signals to/from utilities, other electrical service providers, and market operators. The model incorporates IEC 61850 aspects for generator facilities and OASIS calendaring and EMIX for their energy manager model. It encourages integration of on-site generation, demand response, and energy storage into the main electrical grid.

FACTS (Flexible AC Transmission Systems)
A collection of technologies that manages power flow control and voltage stability on the grid to increase existing transmission capacity and increase or maintain stable voltage operating margins. Products include STATCOMs (Static Compensators) and DVR (Dynamic Voltage Restorers) and may also be deployed on the distribution grid.

Failure
Any electric power failure that threatens the continuity of the bulk electric power supply system to the point that a load reduction action may become necessary. Also known as a hazard.

FAN (Field Area Network)
A network at the distribution level of the grid. Usually proprietary in nature, these networks collect meter data in AMI (Advanced Metering Infrastructure) deployments.

FAST (Feeder Automation System Technologies)
Part of Distribution Automation technologies, it includes remote monitoring and control capabilities on feeders to improve reliability, reduce outages, and extend asset life.

Fault
Any discrete event in the electric system such as a short circuit, a compromised wire, or an intermittent connection.

FC2K (Fuel Cell 2000)
An organization that promotes the commercialization of fuel cells and hydrogen by supplying information about the technology and solutions to a general audience, and provides a portal for R&D, manufacturers, government officials, and others interested in fuel cells. It is part of BTI (Breakthrough Technologies Institute). (http://www.fuelcells.org/)

FCA (Forward Capacity Auction)
Competitive process used in FCMs (Forward Capacity Markets) to procure capacity to meet future forecasts and promote investments through long-term commitments to purchase energy.

FCAPS (Fault, Configuration, Accounting, Performance, Security)
An accepted ITU model for enterprise management. Fault management refers to discovery and management of problems impacting network performance. Configuration management refers to daily operations and coordination of any hardware, firmware or software changes to minimize network performance. Accounting management refers to delivering cost-effective operations and accurate billing. Performance management refers to proactive tasks that anticipate and address potential problems that would impact network performance. Security management refers to protection of the network from physical or cyber security compromises.

FCC (Federal Communications Commission)
The federal agency established by the Communications Act of 1934, charged with regulating interstate and international communications by radio, television, wire, satellite, and cable. It is rebuilding a TAC (Technological Advisory Council) that will provide technical advice to the FCC and make recommendations on a range of communications topics, including Smart Grid. Another office, the WTB (Wireless Transmission Bureau), oversees cellular and PCS phones, pagers, and two-way radios. This bureau also regulates the use of radio spectrum for a variety of applications, including industrial. (http://www.fcc.gov/)

FCC Code of Federal Regulations Sections 15:109 and 15:209
These regulations establish emission limits of electromagnetic interference (EMI) and have would apply to community energy storage.

FCHEA (Fuel Cell and Hydrogen Energy Association)
An industry association dedicated to the commercialization of fuel cells and hydrogen energy technologies. Members include universities, government laboratories and agencies, trade associations, fuel cell materials, components and systems manufacturers, hydrogen producers and fuel distributors, utilities and other end users. (http://www.fchea.org/)

FCL (Fault Current Limiter)
Equipment that detects and dissipates power spikes or surges, allowing normal current to pass through but not fault current. Power spikes can damage electrical equipment or trigger power outages.

FCM (Federation of Canadian Municipalities)
An association that represents municipalities on policy and program matters. Members include Canada's largest cities, small urban and rural communities, and 18 provincial and territorial municipal associations. (http://www.fcm.ca/)

FCM (Forward Capacity Market)
A FERC (Federal Energy Regulatory Commission)-approved wholesale capacity market design intended to procure enough capacity to meet New England's forecasted demand three years in advance through a competitive auction process and to provide a long-term (up to five years) commitment to encourage new investment. A major provision of the FCM settlement is that demand resources, including energy efficiency, distributed generation, real-time demand response, and load management, can qualify as capacity along with supply-side resources. This market design is identified as the next generation to decoupling, and is currently conducted by the ISO-NE.

FCV (Fuel Cell Vehicle)
Any vehicle that takes electricity produced by a fuel cell to feed a storage battery to provide motive power.

FDIR (Fault Detection, Isolation and Repair)
A standard process to identify and fix performance issues.

FDISR (Fault Detection, Isolation, Service Restoration)
Another term for a standard process to identify and fix performance issues.

Federal alternative compliance payment
As defined in the ARRA Section 1306 amending EISA 2007 (Energy Independence and Security Act of 2007), a payment made by utilities to the US federal government that do not meet the minimum number of federal renewable electricity credits as defined in the 2009 ACES legislation.

Federal Electric Utility
A utility that is either owned or financed by the federal government.

Federal Power Act
Legislation created in 1920 and amended in 1935, the Act consists of three parts. The first part incorporated the Federal Water Power Act

administered by the former Federal Power Commission, whose activities were confined almost entirely to licensing non-federal hydroelectric projects. Parts II and III were added with the passage of the Public Utility Act. These parts added regulation of the interstate transmission of electrical energy and wholesale rates in interstate commerce. FERC (Federal Energy Regulatory Commission) now administers this law.

Federal renewable electricity credit
As defined in the ARRA Section 1306 amending EISA 2007 (Energy Independence and Security Act of 2007), a credit equivalent to 1 kWh of renewable electricity as defined in the ACES legislation.

Federal Water Pollution Control Act of 1948
First legislation involving federal government in nation's water sanitation infrastructure. Provided planning, technical services, research and financial assistance by the federal government to state and local governments for sanitary infrastructure. It has become known as the Clean Water Act since 1977, and major amendments were enacted in 1961, 1966, 1970, 1972, 1977, and 1987, including establishment of uniform water quality standards and a Federal Water Pollution Control Administration to set quality standards when states fail to act.

Feeder
An electrical supply line in the utility distribution system that carries power overhead or underground from the substation to transformers.

FEMP (Federal Energy Management Program)
An office within the DOE (Department of Energy) that facilitates the government's implementation of energy management and investment practices for energy security and environmental stewardship through project transaction services, applied technology services, and decision support services. (http://www1.eere.energy.gov/femp/)

Femtocell
A very-low-power 3G mobile phone base station that uses licensed radio spectrum and can also connect via standard broadband DSL or cable service to a mobile service provider. The femtocell encrypts all voice and data sent and received. It has potential as a HAN (Home Area Network) communication device.

Femtogrid
A microgrid scaled to an individual home unit that includes self-contained generation, distribution, energy storage, and energy management software with a seamless and synchronized connection to a utility power system, but can operate independently from the grid for a defined period of time.

FERC (Federal Energy Regulatory Commission)
A federal agency that has oversight of interstate electricity sales, wholesale electricity rates, oil pipeline rates, natural gas pricing and gas pipeline certification, and hydroelectric licensing. FERC is responsible for publishing standards on Smart Grid interoperability to ensure that devices and technologies deployed across the grid work together. It is an independent regulatory agency within the DOE (Department of Energy) and is the successor to the Federal Power Commission. (http://www.ferc.gov/)

FERC Order 755
This rule, issued in February 2013, approves the Regional Reliability Standard PRC-006 NPCC-01, which regards Automatic Underfrequency Load Shedding (UFLS). The standard is aimed at ensuring the development of an effective UFLS program in order to maintain the integrity of the grid when it is under pressure from heightened demand. A facet of this program is the development of fast responding generation sources to meet demand peaks. It applies to generator owners, planning coordinators, distribution providers, and transmission owners in the Northeast Power Coordinating Council Region.

FERC Order 779
A reliability standard requiring transmission operators and coordinators to develop processes and policies that address risks and mitigate impacts to bulk power infrastructure posed by GMDs (Geomagnetic Disturbances) such as solar flares.

FERC Order 784
This rule expands options for energy storage on a grid by reforming ancillary service regulations. It allows fast batteries, flow batteries, and mechanical flywheels to compete against slower reacting gas or coal fired plants in the ancillary service market. The new accounting rules introduced in this order allow utility companies to achieve rate recovery for energy storage equipment.

Ferroresonance
A distribution system transient resulting from switching operations, ground faults, or even lightning. It appears as very highly sustained overvoltages concurrent with high levels of harmonic distortion, and can damage power equipment.

FES (Flywheel Energy Storage)
A non-electrochemical technology that stores and releases energy as needed. FES may be used to address peak demand.

Fiber optics
A wired communications medium made of glass with high speed and high bandwidth capabilities. It is more resistant to electrical "noise" than other wired or wireless media, and is robust over long distances without requiring boosts in signaling.

Field capacitor bank
Equipment in the distribution network that controls voltage.

File rate schedule
The documented electric rates and all terms and conditions for a utility's pricing structure that is approved by the regulatory authority with oversight of that utility.

Filing
Any written documentation that is submitted to a utility commission.

Filtration
A process in surface water treatment plants that uses filters, often made from sand, activated carbon or anthracite, to remove small particulates from water.

Final order
A final ruling by FERC (Federal Energy Regulatory Commission) that closes the matter under discussion.

FInES (Future Internet Enterprise Systems)
A European research initiative that is part of FP7 and is focused on the vision of the future Internet. It combines two previously formed research groups - the Enterprise Interoperability, and Collaboration and Digital Ecosystems, also known as clusters, to build on their foundation of knowledge to define future enterprise systems needed to support the future Internet. (http://cordis.europa.eu/fp7/ict/enet/ei_en.html)

FIPPs (Fair Information Practice Principles)
Five core principles of privacy protection identified as: 1) Notice/Awareness, 2) Choice/Consent, 3) Access/Participation, 4) Integrity/Security, and 5) Enforcement/Redress. These principles are identified as important governing principles for information collected by smart meters. (http://www.ftc.gov/reports/privacy3/fairinfo.shtm)

FIPS (Federal Information Processing Standards)
NIST (National Institute of Standards and Technology) standards and guidelines for federal computer systems, as directed by the Information Technology Management Reform Act (Public Law 104-106) and approved by the Secretary of Commerce. FIPS are developed for interoperability and security situations when there are no acceptable industry standards. These standards do not apply to national security systems. (http://www.itl.nist.gov/fipspubs/geninfo.htm)

FIPS 140-3
Part of the FIPS standards group, this is the proposed revision to FIPS 140-2. It is used to accredit cryptographic modules, and can be used for data security within the Smart Grid domain. It reintroduces firmware cryptographic technologies and limits software cryptographic modules to Security Level 2, and strengthens integrity testing.

Firm power
The power or capacity that would be available at all times during the period covered by a guaranteed commitment to deliver, regardless of conditions.

Firming
The application of energy storage or blended sources of energy to provide predictability to intermittent renewables and resolve low voltage ride through events.

Firmware
Software that is programmed in a hardware device's read-only memory.

First Fuel
A concept that identifies energy efficiency as the first fuel to set priorities about design, deployment, and use of any devices or materials that consume electricity or contribute to its consumption.

FiT (Feed-in Tariff)
An energysupply policy that encourages new renewable power generation and attempts to provide investor certainty with guarantees of payments in dollars per kWh for the full output of the system for a guaranteed period of time.

Fixed costs
A utility and regulatory term for some of the costs to provide electricity to ratepayers. It includes investment costs and the interest on debt and return on equity for those investments; unavoidable maintenance costs for generation, transmission, and distribution assets; and payroll.

FIXML
Also known as Financial Information Exchange, FIXML is an ASCII based electronic communications protocol for the real time exchange of information specifically related to securities transactions and large scale financial data.

Flash power plant
A type of geothermal power plant that takes heated water under pressure and separates it in a surface vessel into steam and hot water. The steam is delivered to a turbine that in turn powers a generator. The liquid is returned to the reservoir. The other types of geothermal power plants are binary power plant, dry steam power plant, and flash/binary combined cycle.

Flash/combined cycle power plant
A type of geothermal power plant using a combination of flash and binary technology. The geothermal water that converts to steam is converted to electricity with a backpressure steam turbine and the low-pressure steam exiting that turbine is condensed in a binary system. The other types of geothermal power plants are binary power plant, dry steam power plant, and flash power plant.

Flat and meter rate schedule
A two-part electric rate schedule that consists of a service charge and a price for the energy consumed in a set time frame.

Flexible demand
Devices that can be added as load to the grid (generally distribution) at any time, but preferably off-peak or to help address HVRT (High Voltage Ride Through).

FLISR (Fault Location, Isolation, Service Restoration)
Another term for a standard process to identify and fix performance issues.

Flow cell or battery
A battery technology that uses an active element in a liquid electrolyte that is pumped through a membrane similar to a fuel cell to produce an electrical current. The size and number of membranes determine power rating, and the runtime, or hours of discharge, is based on the gallons of electrolyte pumped through the membranes. Pumping in one direction produces power from the battery, and reversing the flow charges the system. Common characteristics of this energy storage technology are extremely high cycle life and short charge times without extended charge-balancing requirements. Vanadium redox and Zinc-Bromide are two examples of flow cells. It is suitable energy storage for distributed generation.

Flow limiting valve
A valve that limits the flow in the pipe to the established design flow for the system, also known as a differential pressure-activated flow control valve.

Flow meter
An instrument indicating the velocity and/or volume of a fluid like water.

Fluid hammer
The sudden start/stop of water flowing in a pipe, resulting in a loud thudding sound. Also known as water hammer, it can be destructive to sensors.

Flywheel
A mechanical device that stores rotational energy. Some important characteristics of flywheels are low maintenance, long life (20 years or tens of thousands of deep cycles), and benign materials (no chemicals). Flywheels can perform for short-term ride-through and long-term storage. Traditionally used in aerospace and UPS applications, it is now finding use in telecom operations. Megawatts for minutes or hours are stored through a flywheel farm design.

FM (Frequency modulation)
A technique for transmitting information via radio carrier waves in which the frequency of the carrier waveform is varied.

Forced outage
The suspension or shutdown of generation, transmission, or distribution components due to an emergency, such as a weather disaster or an unanticipated equipment failure.

FP7 (Seventh Framework Programme)
An oversight office that bundles all research-related EU initiatives together with the objective of building the most competitive knowledge-based economy. Smart Grid research and other energy research initiatives are included here. (http://cordis.europa.eu/fp7/understand_en.html)

Frame Relay
A wide area network service that transmits packets between local area networks or nodes. It operates at lower speeds than ATM service but does not require dedicated connections. It also offers prioritization of packets.

FRCC (Florida Reliability Coordinating Council, Inc.)
A not-for-profit entity that serves as a reliability region with delegated authority from NERC (North America Electric Reliability Corporation) to propose and enforce reliability standards within peninsular Florida east of the Apalachicola River. Areas west of the river are within the SERC Region. The FRCC ensures and enhances the reliability and adequacy of bulk electricity supply in its territory. The entire FRCC Region is within the Eastern Interconnection. It is one of the eight NERC Regional Entities or Reliability Regions. (https://www.frcc.com/AboutUs/default.aspx)

Free rider
An enrollee in a DR (Demand Response) program or other program designed to encourage specific energy behaviors who has to make little or no effort to comply with program requirements to enjoy the financial rewards of participation.

FREEDM Systems Center
A Gen-III Engineering Research Center founded by the National Science Foundation in 2008. It includes utilities and national laboratories to develop technology to revolutionize the nation's power grid and speed the introduction of renewable electric-energy technologies. Its objectives include delivering fundamental breakthrough technology in energy storage and power semiconductor devices; and developing long-term partnerships with middle and high schools to build engineering enrollments. (http://www.freedm.ncsu.edu/)

FreedomCAR and Fuel Partnership
A public/private partnership program in the DOE (Department of Energy) that examines high-risk research to develop component and infrastructure technologies to reduce the dependence of transportation system on imported oil and minimize harmful vehicle emissions. (http://www1.eere.energy.gov/vehiclesandfuels/about/partnerships/freedomcar/index.html)

Frequency
A term that describes the number of radio waves that pass a point in space per second, measured in Hertz (Hz). For the US power grid, the system frequency is 60 Hz.

Frequency regulation
The portion of power kept on reserve by utilities or Balancing Authorities to maintain a scheduled frequency.

Fresh water
Water that contains less than 1,000 ppm (parts per million) of dissolved solids of any type.

FRT (Fault Ride-Through)
The ability of wind turbines to continue operations during grid disturbances such as voltage drops using capacitor banks and battery backup. It is usually required for grid integration. Also known as Low Voltage Ride-Through or LVRT.

FSO (Free Space Optical)
An LoS (Line of Sight) technology that transmits a modulated beam of visible or infrared light in the TeraHertz portion of the spectrum through the atmosphere for broadband communications. The beam is transmitted through space instead of along a cable. It does not require spectrum licensing. Important characteristics are its high level of security and its severe transmission impacts from fog and turbulence caused by air pockets.

FTC (Federal Trade Commission)
US Federal agency that has oversight of business practices of all types of corporations except financial institutions and common carriers. It works to prevent unfair, anti-competitive, and deceptive business practices and to protect consumers through information and enforcement. It provides

policies and guidance to businesses and consumers regarding data privacy and data security. (http://www.ftc.gov/)

FTR (Financial Transmission Rights)
Financial instruments that give their holders (generators or loads) the difference between the locational price at the point where power is placed in the market and the locational price at the point where power is withdrawn from the market. The holders receive the value of congestion as established by the locational price difference. FTRs are used like insurance to hedge the costs associated with transmission congestion, and helps facilitate competitive open transmission access.

FTU (Feeder Terminal Unit)
A remote terminal unit along distribution feeders that communicates a local process to a central system. It is also the control equipment of many automated load break switched, or auto-reclosers for distribution automation. The primary functions of an FTU are remote operation of switches, status monitoring of switches, remote measurements of electric values, and fault detection and protection.

FUCO (Foreign Utility Company)
A term used in PUHCA 2005 that identifies any company that owns or operates or derives income from facilities that are used for the generation, transmission, or distribution of electric energy for sale or the distribution at retail of natural or manufactured gas for heat, light, or power outside of the United States.

Fuel cell
A battery that converts chemical energy to electrical energy. It is different from typical batteries because the active materials (fuel and an oxidant like oxygen) are supplied from outside, not within its structure. Fuel cells may be used for on-grid or off-grid power generation. Fuel cells run on hydrogen, which can be extracted from water, methanol, ethanol, natural gas, ammonia, and renewables like biomass. Fuels containing hydrogen

require a "fuel reformer" that extracts the hydrogen. Although much focus has been on transportation applications, certain fuel cell technologies are researched for small- to large-scale power generation on or off the Smart Grid, including SOFC (Solid Oxide Fuel Cell), MCFC (Molten-Carbonate Fuel Cell), and PAFC (Phosphoric-Acid Fuel Cell).

Future of Privacy Forum
A nonprofit think tank that seeks to advance responsible data practices. The forum is led by Internet privacy experts and includes an advisory board that comprises leading figures from industry, academia, law, and advocacy groups. (http://www.futureofprivacy.org/)

FWEE (Foundation for Water and Energy Education)
An organization comprised of public utility districts and a federal power agency focused on water use in the Northwest, with a goal of ensuring that water remains a renewable resource. It provides information on hydroelectric power facilities. (http://www.fwee.org/)

G

G.hn
G.hn is another name for a suite of standards. ITU-T Recommendation G.9960 is a MAC/PHY (Media Access Control/PHYsical) standard for high-speed IP-based home networking that spans coaxial cable, electrical wiring and phone lines. Transmission speeds vary for the medium. For coaxial cable, the maximum speed is 800Mbps, for power lines, between 200Mbps and 400Mbps, and copper phone lines will support 200Mbps or more. G.9961 addresses the Data Link Layer (DLL), and G.9972 addresses the coexistence functions. G.9970 covers the transport architecture or Layer 3 for home networks.

G.hnem
A set of proposed ITU-T standards that support communications over low and medium voltage lines, through low and medium voltage transformers. G.9955 addresses PHY and system architecture, and G.9956 covers the DLL (Data Link Layer) for powerline communications in the 9-500 KHz range. These complement G.hn with a focus on networked energy management applications. They are considered part of the G.9960 family of recommendations.

G3-PLC™
An open and non-proprietary narrowband PLC (Power Line Communications) protocol that is based on OFDM (Orthogonal Frequency Division Multiplexing). It coexists with IEC 61334, IEEE P1901 and ITU G.hn.

G3-PLC Alliance
An industry association that promotes G3-PLC in standards discussions, advocates testing frameworks for interoperability, and delivers education. (http://www.g3-plc.com/index.html)

Galvin Electricity Initiative
An organization promoting the perfect power system, which is robust and resilient, is able to withstand natural and weather-related disasters,

mitigates the potential damage caused by terrorist attack, provides affordable electricity to all consumers, and allows consumers to control their own energy use to the extent they choose. Their reports describe existing technology, assessments of technology innovations, and functional specifications. Part of the promotion involves prototypes of this power system to demonstrate its advantages and efficacy of its plans. (http://www.galvinpower.org/)

Gamification
The application of game thinking and game mechanics in non-traditional environments to educate participants, increase interaction, and develop creative responses to challenges through fun, rewards, and social connections.

Gate valve
A water system valve that operates either fully open or fully closed. They prevent fluid hammer, which can be destructive to pipes.

GEA (Geothermal Energy Association)
A USA trade association of companies who promote use of geothermal energy for electrical power generation and direct-heat uses. It encourages R&D to improve geothermal technologies, provides export assistance for geothermal goods and services, compiles statistical data about the geothermal industry, and conducts education and outreach. (http://www.geo-energy.org/)

GeoEVSE Forum
This is a partnership of the U.S. Department of Energy, Google, and over 80 organizations working for electric vehicle (EV) deployment. The goal of the program is to be a primary data source for mapping electric vehicle charging (or EVSE) stations. With this data, EV drivers will be able to easily locate charging stations to "refuel" their electric vehicles.

Generating station
A station with electric generators and auxiliary equipment that converts mechanical, chemical, or nuclear energy into electric energy.

Generating unit
Generators, reactors, boilers, combustion turbines, and other prime movers that produce electric power.

Generation
The process of producing electric energy from a non-renewable or renewable source, measured in kilowatt hours or megawatt hours.

Generation company
An entity that owns or operates generating plants or interacts with the short-term market on behalf of plant owners.

Generator capacity
The maximum output that generating equipment can supply to system load, adjusted for ambient conditions. It is usually identified in megawatts.

Generator nameplate capacity
The maximum rated output of a generator under specific conditions designated by the manufacturer. Also known as nameplate capacity.

GENIVI® Alliance
A nonprofit industry alliance promoting the broad adoption of an In-Vehicle Infotainment (IVI) open-source development platform. IVI includes music, news, navigation and location services, telephony, and Internet services. The association helps development by alignment to automotive OEM requirements through specifications, reference implementations, and certification programs. (http://www.genivi.org/)

Geoexchange
A heat transfer or exchange between the ground and the building that is accomplished by using standard pump and compressor technology. Also known as earth energy systems or geothermal heat pump systems.

Geofluid
As defined in the ARRA Section 1306 amending EISA 2007 (Energy Independence and Security Act of 2007), any fluid except a fossil fuel that is used to extract underground thermal energy for direct use or electricity generation.

Geothermal
An underground heat source that can be accessed for direct use or electric power generation. High temperature (greater than 150 degrees Celsius) geothermal resources are used for electricity generation. Resources below 150 degrees Celsius are used for direct heat applications. Electricity generation requires conventional steam turbines or binary plants.

GFCI (Ground Fault Circuit Interrupter)
A device that monitors a household electrical grid for ground faults. If a GFCI device senses any loss of current in a circuit, it automatically switches off power to that circuit.

GHG (Greenhouse Gas)
The atmospheric gases that are the cause of global warming. Those gases, such as water vapor, carbon dioxide, nitrous oxide, methane, HFC (hydrofluorocarbons), PFC (perfluorocarbons), and sulfur hexafluoride, are transparent to solar radiation but opaque to long-wave or infrared radiation, thus preventing long-wave radiant energy from leaving Earth's atmosphere. The net effect traps absorbed radiation and warms the planet's surface.

GIA (Generator Interconnection Agreement)
A contract that outlines the procedures for connecting a generation asset to the power grid. There are standard agreements defined by FERC (Federal Energy Regulatory Commission) for large generation assets over 20 megawatts and small generation assets less than 20 megawatts.

GID (Generic Interface Definition)
Part 4 of IEC61970 that specifies component interfaces to facilitate integration with other components that may have been developed by other manufacturers.

GIS (Geographic Information System)
A system that integrates hardware, software, and data for capturing, managing, analyzing, and displaying all forms of geographically referenced information. It is used by utilities for power, facilities, and asset management and may be integrated to existing applications like customer service and SCADA to enhance operations.

GITA (Geospatial Information & Technology Association)
A nonprofit educational association serving the global geospatial community. It advocates for anyone using geospatial technology to help operate, maintain, and protect the utilities, telecom, and public sector infrastructure. It delivers education and professional best practices. This group has an initiative called CIP-ER (Critical Infrastructure Protection – Emergency Response) that focuses on use of geospatial applications. It was formerly known as AM/FM International (Automated Mapping/Facilities Management). (http://www.gita.org/)

GLD (Guaranteed Load Drop)
A demand response program that reduces electricity consumption by a specific amount of kilowatts for a specific time frame. It measures actual usage in that timeframe to typical usage without a curtailment agreement.

This type of agreement offered by curtailment service providers and is available in PJM territory.

Global Smart Grid Federation
An international federation established to share ideas and best practices about Smart Grid technology development, consumer engagement, and capacity building. It also facilitates public-private dialog about Smart Grid policy implementations based on ongoing deployments. Members include the GridWise Alliance, India Smart Grid Forum, Japan Smart Community Alliance, Korean Smart Grid Association, Smart Grid Australia, SmartGridIreland, and the Canadian Smart Grid Alliance. (http://www.globalsmartgridfederation.org/)

Global warming
An increase in the near surface temperature of the Earth. Global warming refers to the warming that is occurring as a result of increased emissions of greenhouse gases caused by human activity.

GMS (Generation Management System)
Integrated application software suitable for monitoring, analysis, and control of electric power generation resources. It allows utility asset managers to schedule the delivery of generation capacity from each plant according to changing demand from contracted markets. The software also lets users monitor performance and real-time costs, schedule transactions, and interface to other systems like SCADA (Supervisory Control And Data Acquisition) and EMS (Energy Management Systems).

GMR-1 3G (Geostationary Earth Orbit Mobile Radio-1 3G)
ETSI (European Telecommunications Standards Institute) standards that specify interface guidelines for circuit and packet switch networks from both GSM and 3GPP with geostationary satellite communications.

GO 131-D (General Order 131-D)
This CPUC rule addresses planning and construction in California of electric G/T/D facilities, lines, and substation facilities between 50 kilovolt (kV) and 200 kV. For purposes of this GO, a transmission line is a line that operates at or above 200 kV. A power line operates between 50 and 200 kV. A distribution line that operates under 50 kV. Any facility over 50kV requires CPUC (California Public Utilities Commission) approval.

GO 15
A global industry body made up of the sixteen largest electric grid operators around the world, to address power system security. (http://www.go15.org)

GOOSE (Generic Object-Oriented Substation Event)
Data communications using Ethernet to transmit substation commands, alarms, and status updates as event messages that are targeted to specific IEDs (Intelligent Electronic Devices) or receivers. GOOSE provides interoperability of devices in substations as defined by IEC 61850.

GPM (Gallons/minute)
A common metric for water consumption.

GPR (Ground Potential Rise)
A current flow in the earth caused by electrical faults at electrical substations, power plants, or high-voltage transmission lines. Short-circuit current flows through the plant structure and equipment and into the grounding electrode at station. Factors that determine the hazard in this current flow include soil conditions or the amount of current entering the earth.

GPRS (General Packet Radio Service)
A packet-switched technology that enables wireless data communications. GPRS is informally known as 2.5G technology, better than 2G, but not as

fast as 3G. Data rates range from 56 to 114 Kbps and services include always-on Internet connections for mobile phone and computer users. GPRS supports IP and X.25 protocols.

GPS (Global Positioning System)
A satellite-based navigation system that delivers reliable positioning, navigation, and timing services. It consists of satellites, control and monitoring stations, and GPS receivers. Each receiver delivers latitude, longitude, altitude, and time.

GRAPES (Grid Connected Advanced Power Electronics Systems)
A basic research initiative started by the NSF (National Science Foundation) and two universities. The mission is to accelerate the adoption and use of power electronics in the electric grid to improve system stability, flexibility, robustness, and economy. (http://www.grapes.uark.edu/)

Gray water
Wastewater that is produced from clothes washers, bath sinks, bathtubs, and showers. It can be used for irrigation purposes in some jurisdictions.

GRC (Geothermal Resources Council)
A nonprofit, educational association for the international geothermal community. It encourages international development of geothermal resources, promotes research, exploration and development of geothermal energy in ways compatible with the environment, provides a forum for the world geothermal community, and cooperates with national and international academic institutions, industry and government agencies to encourage economically and environmentally sound development and utilization of geothermal resources. (http://www.geothermal.org/)

GRDA (Geothermal Resources Development Account)
A California grant fund that promotes the development of new or existing geothermal resources and technologies. It is the only fund of its type in the

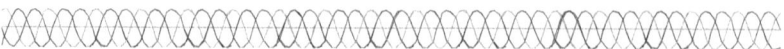

USA. It is funded from geothermal royalty revenues paid to the federal government by geothermal developers for leases on federal land in California. Eligible private entities and local jurisdictions can qualify for assistance in geothermal research, development, demonstration, commercialization, planning, mitigation and environmental enhancement projects. (http://www.energy.ca.gov/geothermal/grda.html)

Green Bank
A proposed public financial institution that would work in partnership with private sector entities to open credit markets and investments in large-scale renewable energy and energy efficiency technologies and projects.

Green Building Initiative
A nonprofit that promotes the Green Globe certification.

Green Circuits
An Electric Power Research Institute (EPRI) transmission and distribution efficiency initiative. It is a collaborative project with a number of international utilities to study and demonstrate investments in infrastructure that improve overall energy efficiency and energy conservation.

Green eMotion
This is a project to expand the use of electric vehicles (also known as electromobility) in Europe in an effort to reduce vehicle generated CO_2 emissions. It is part of the European Green Cars Initiative, which is a part of the European Economic Recovery Plan.

Green Globe
An environmental assessment, education and rating system.

Green Grid
A global consortium of computer, chip, and other manufacturers and hosting services dedicated to developing and promoting energy efficiency for data centers. It creates standards, measurement practices, and processes, and promotes energy efficiency standards, processes, and technologies. (http://www.thegreengrid.org/about-the-green-grid)

Green pricing
Specialized pricing and voluntary programs to purchase renewable energy and fund its development through payments on customer utility bills.

GreenTouch
An industry consortium of Information and Communications Technology (ICT) industries, academic institutions, and non-governmental research experts focused on reducing energy requirements and carbon emissions through communications and data network, platform and device innovations. (http://www.greentouch.org)

Grid
The lines and equipment that transport electric energy between generation facilities and points of consumption. At high voltage, it is the transmission grid. At low voltage, typically defined as below 33 kV (kilovolts) for the US electrical system, it is the distribution grid.

Grid 2.0
A term used to describe the expansion of bi-directional communications and intelligent devices to automate distribution grids and substations. Grid 1.0 focused on smart meters and FANS (Field Area Networks).

Grid parity
The price point at which renewable energy is competitive with traditional fuel sources, such as fossil fuels and nuclear power. (Editor's Note: In the USA, fossil-based energy sources receive significant subsidies, and there are

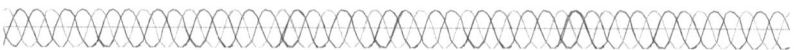

compelling arguments that elimination of these subsidies to mature and profitable industries, plus addition of true carbon costs, would significantly change the grid parity equation.)

Grid-scale energy storage
Storage that is 25kW or higher. Below this level is building-scale storage.

Grid stabilization
A process in which electricity stored in electrochemical or non-electrochemical devices is pumped onto the grid for a short time, from a few seconds to under an hour, to match grid demand and supply to make generators run more efficiently or to stabilize grid signal frequency. Also known as Grid support.

GridApp (Advanced Grid Applications Consortium)
A consortium of utilities working to provide a fast-track process for engineering development, demonstrations, and validation of technologies for a Smart Grid. Concurrent Technologies Corporation is the GridApp organizer and provides management and technology transition support to GridApp, with funding from the DOE's (Department of Energy) OE and utility members. (http://gridapp.org)

GridWise®
A registered trademark owned by the Pacific Northwest National Laboratory that was developed with the DOE (Department of Energy). The GridWise brand has been used by organizations with the permission of PNNL for efforts that are not necessarily connected to each other.

GridWise® Alliance
Founded in 2003, a coalition of public and private stakeholders that advocates for a smarter grid for the public good. The Alliance facilitates the effective collaboration among stakeholders to promote, educate, and advocate for the adoption of innovative Smart Grid solutions. These

solutions will achieve economic and environmental benefits for customers, communities, shareholders, and society. GridWise Alliance members include utilities, IT companies, equipment vendors, new technology providers, and academic institutions. (http://www.gridwise.org)

Gross savings
A regulatory term that describes energy efficiency program savings as one of three levels. Gross savings means ex ante estimates adjusted for the actual program conditions or results. The other two measures are ex ante savings and net savings.

Ground potential rise
An electrical fault condition found at substations, power plants, transmission lines, or telecom facilities. Ground has finite resistance to current, so any increase from a short circuit or lightening strike creates a potential rise between any two points, and can cause step and touch hazards.

Groundwater
Water that is located below ground in aquifers, springs, and wells.

Groundwater Foundation
A nonprofit organization focused on education about clean and sustainable groundwater issues. It provides community-based programs to involve individuals, communities, and both private and public entities in groundwater conservation and protection activities. (http://www.groundwater.org/)

GS OSG 001 (Group Specification Open Smart Grid 001)
An application layer protocol produced by the ETSI Open Smart Grid Industry Specification Group, which is part of ESNA and the NES architecture.

GSDI (Global Spatial Data Infrastructure) Association
An international association of organizations, agencies, firms, and individuals to promote international cooperation and data sharing in support of local, national and international spatial data infrastructure developments that will allow nations and their citizens to better address social, economic, and environmental issues. It has specific applications to disaster recovery from natural and human-caused events and could be an important component of smart cities. (http://www.gsdi.org/)

GSI (Green Storage Initiative)
A program within SNIA (Storage Networking Industry Association) to advance energy efficiency and conservation in all networked storage technologies and minimize the environmental impact of data storage operations. Its GSI Technical Work Group produced an SNIA Green Storage Power Measurement Technical Specification working draft in early 2009 to help companies assess power utilization of storage products. (http://www.snia.org/forums/green/)

GSM (Global System for Mobile Communication™)
A 2G broadband wireless technology that uses a variation of TDAM (Time Division Multiple Access). It operates in the 900MHz and 1.8GHz bands in Europe and the 1.9GHz and 850MHz bands in the USA. The 850MHz band is also used for GSM in Australia, Canada, and many South American countries.

GTI (Gas Technology Institute)
A not-for-profit industry association focused on research and development regarding natural gas energy supply, delivery, and end use. (http://www.gastechnology.org)

GW (Gigawatt)
One billion watts or one thousand megawatts.

GWAC (GridWise® Architecture Council)
An industry group formed by the DOE (Department of Energy) to promote and enable interoperability among the entities that interact with the electric grid, through proposals and tools to guide development in interoperability standards. The DOE is providing limited financial and logistical support to the Architecture Council. GWAC sees the intelligence emerging at the fringes of the traditional electric grid system as transformational to the electric power business and its operations. Thirteen members are selected to represent the full spectrum of industry and academia with complementary experience relevant to planning for future grid technologies and structures. NIST (National Institute of Standards and Technology) has joined with GWAC to form DEWGs (Domain Expert Working Groups) to explore Smart Grid interoperability issues. GWAC coordinates with a number of organizations in related fields, including the intelligent building community, NRECA's (National Rural Electric Cooperative Association) Multispeak initiative, and architecture efforts in the ISO/RTO area. (http://www.gridwiseac.org/)

GWh (Gigawatt hour)
One billion watt hours.

GWPC (Groundwater Protection Council)
A nonprofit organization of state ground water regulatory agencies focused on the protection of the nation's ground water supplies. It promotes best management practices and legislation for ground water protection. (http://www.gwpc.org/home/GWPC_Home.dwt)

H

H2G (Home to Grid)
An acronym describing either the connection of residential homes to the Smart Grid or a working group within NIST (National Institute of Standards and Technology) focused on this subject. (www.nist.gov/smartgrid/)

H2G-enabled
Homes that are capable of sending power back to the grid. The home and the grid it is plugged into must have the following: communications, price signals, charge interruption features, and billing support.

HAC (Hot Aisle Containment)
A data center practice of physically separating hot exhaust air and cold air by covering a hot aisle and delivering cool ambient air to the server rack air inlets.

HAN (Home Area Network)
A network of energy management devices, digital consumer electronics, signal-controlled or enabled appliances, and applications within a home environment that is on the home side of the electric meter. It is similar to a home-based LAN, but it connects more than personal electronics like computers, printers, and TVs. HAN specifications include OSHAN, ZigBee, HomePlug, Z-Wave and Wireless M-Bus (a wireless variant of M-Bus).

Hard water
Water that contains high levels of calcium, magnesium, and other minerals.

Harmonics
A power disturbance generally caused by loads coming on line from office equipment, consumer electronics, and lighting.

HART® (Highway Addressable Remote Transducer)
A bi-directional communication protocol enabling data exchange in industrial process measurement and control instrumentation and systems. Initially developed for wired communications, it now covers wireless communications.

HART Communication Foundation
An international not-for-profit membership organization that sets the standards and owns the technology for the HART communication protocol. (http://www.hartcomm.org/)

HCA (High Consequence Area)
Urban, high-population areas that have pipelines (natural gas, water, or other materials) sited there or nearby. It can also include towns, villages or other residential or commercial areas like hospitals or schools, or locations where people congregate.

HD-PLC Alliance (High Definition Power Line Communications)
A global industry organization promoting PLC technology in broadband networks through standards, compatibility testing, and information sharing. (http://www.hd-plc.org/)

HDD (Heating degree-days)
A measure of how cold a location is over a period of time relative to a base temperature, usually 65 degrees Fahrenheit. The measure is computed for each day by subtracting the average of the day's high and low temperatures from the base temperature with negative values set equal to zero. Each day's heating degree-days are summed to create a heating degree-day measure for a specified reference period. Heating degree-days are used in energy analysis as an indicator of space heating energy requirements or use.

Headgate
The upstream gate that controls water flow into irrigation canals and ditches.

Heat pump
Heating and/or cooling equipment that draws heat into a building from outside during the heating season and, during the cooling season, ejects heat from the building to the outside. Heat pumps are vapor-compression refrigeration systems whose indoor/outdoor coils are used reversibly as condensers or evaporators, depending on the need for heating or cooling.

Heating equipment
Any equipment designed and/or specifically used for heating ambient air in an enclosed space. It includes central warm air furnace, heat pump, plug-in or built-in room heater, boiler for steam or hot water heating system, heating stove, and fireplace.

Heating intensity
The ratio of space-heating consumption or expenditures to square footage of heated floor space and heating degree-days. The base is 65 degrees Fahrenheit. This ratio provides a way of comparing different types of housing units and households by controlling for differences in housing unit size and weather conditions. The square footage of heated floor space is based on the measurements of the floor space that is heated. The ratio is calculated on a weighted, aggregate basis according to the following formula: Heating Intensity = Btu for Space Heating / (Heated Square Feet * Heating Degree-Days).

Hedging
The buying and selling of futures contracts so as to protect energy traders from unexpected or adverse price fluctuations.

Hedging contracts
Contracts that establish future prices and quantities of electricity independent of the short-term market. Derivatives may be used for this purpose.

HEMS (Home Energy Management System)
Software that reports and/or manages electricity consumption in home appliances and electronics, and can respond to electricity pricing and usage information. It may include motivators to reduce consumption and improve energy efficiency based on neighborhood and/or demographic information. HEMS solutions may include hardware to provide information or control of appliances, thermostats, and other electricity-using devices in homes. It may include micro-generation management, energy storage, and smart-charging capabilities.

HERS (Home Energy Rating System)
This is a nationally recognized scoring system that generates an index score that gives prospective buyers and homeowners a detailed snapshot of how homes rank in terms of energy efficiency. The index also generates a report that identifies the inefficiencies of the home, so owners can take action to improve their HERS index score.

Hertz
A radio frequency metric that measures in cycles per second as kilohertz (KHz), megahertz (MHz) or gigahertz (GHz). Higher frequencies result in shorter wavelengths.

HES Suite (Home Energy Saver)
A simulation model developed by the Environmental Energy Technologies Division at LBNL (Lawrence Berkeley National Laboratory) that offers online tools that allow homeowners and energy professionals to monitor and evaluate home energy use. It includes recommendations based on both empirical usage data and predicted energy use to save energy. With

accurate data, the HES simulation can predict future home energy use within 1% of actual consumption.

HES (Hybrid Energy Storage)
Hybrid energy storage includes combinations of power and energy devices such as battery plus capacitor, battery plus flywheel, or battery plus battery of different types to enhance the integration of existing power systems with energy storage.

HEV (Hybrid Electric Vehicle)
Any vehicle that has both an internal combustion engine and an electric motor to provide motive power. The electric motor is charged by a battery that is continuously recharged by a generator driven by the internal combustion engine.

High speed relay
A category of switches with activating mechanisms designed for short operate time, short release time, or both.

HMI (Human Machine Interface)
The displays and controls that allow humans to operate equipment and review data. This term is often used in conjunction with devices and substation automation.

Holding company
A company that confines its activities to owning stock in and supervising management of other companies. The SEC (Securities and Exchange Commission) administers the Public Utility Holding Company Act of 1935 and defines a holding company as "a company which directly or indirectly owns, controls or holds 10 percent or more of the outstanding voting securities of a holding company" (15 USC 79b, par. A (7)).

HomeGrid Forum
A global nonprofit industry association promoting the ITU's (International Telecommunications Union) G.hn standard for home networking. It has three work groups: a g.hn Contribution workgroup, a Compliance and Interoperability workgroup and a Marketing workgroup to establish industries' technical requirements, ensure interoperability, and brand and market HomeGrid Certified Products. It is a companion organization to the ITU and members include global ICT companies and consumer electronics manufacturers and retailers.

HomePlug® Powerline Alliance
An industry trade group focused on promoting and certifying interoperable, standards-based HomePlug networks and products using powerline communications. (http://www.homeplug.org/)

HomePNA® Alliance
An industry group of technology vendors that develop, promote and support home networking solutions based on internationally recognized, open and interoperable HomePNA standards documented by the ITU (International Telecommunications Union). (http://www.homepna.org/)

HOMER (originally meant Hybrid Optimization Model for Electric Renewables)
A free software program from NREL (National Renewable Energy Laboratory). It simplifies the task of evaluating design options for both off-grid and grid-connected power systems for remote, standalone, and distributed generation applications. The economic and technical feasibility of a large number of technology options can be evaluated. HOMER models both conventional and renewable energy technologies. NREL provides classroom and individualized training in the use of HOMER. (https://analysis.nrel.gov/homer/)

Hometown Connections
An APPA (American Public Power Association) subsidiary that negotiates national group pricing and service arrangements for its members in utility operations, business planning, and retail services. (www.hometownconnections.com)

HRSG (Heat Recovery Steam Generator)
A type of heat exchanger that recovers energy from a hot waste gas stream. This waste heat or steam can then be reused in cogeneration processes or converted to electricity by powering a steam turbine.

HSDPA (High-Speed Download Packet Access)
See HSPA.

HSPA (High-Speed Packet Access)
A widely deployed mobile broadband technology that refers to deployments of Download (HSDPA) and Upload (HSUPA) network technologies and HSPA Evolved. Download speeds range from 3.6 to 14 Mbps and upload speeds range from 0.4 to 5.7 Mbps. It is a 3G technology.

HSPD-5 (Homeland Security Policy Directive 5)
A directive that developed the National Response Plan (NRP) and the National Incident Management System (NIMS) to manage domestic incidents. Emergency Support Function #12 in the NRP targets incidents with the national energy infrastructure to restore damaged energy systems and components. Although energy facility owners are responsible for restoration of normal operations, the DOE (Department of Energy) is the primary federal agency to help in restoration activities.

HSPD-7 (Homeland Security Policy Directive 7) A directive that established a national policy for Federal departments and agencies to identify and prioritize critical infrastructure and key resources and to protect them from terrorist attacks.

HSPD-8 (Homeland Security Policy Directive 8)
A directive replaced with Presidential Policy Directive 8 (PPD-8) that focuses on improvements to national preparedness through attention to systemic security and resiliency to security threats.

HSUPA (High-Speed Upload Packet Access)
See HSPA.

HSWA (Federal Hazardous and Solid Waste Amendments)
Legislation passed in 1984 that amended RCRA and included increased EPA enforcement authority; a broad underground storage tank program; and reduction of land-based hazardous waste disposal.
(http://www.epa.gov/lawsregs/laws/rcra.html)

HTS (High-Temperature Superconductivity)
Technology used for high-voltage transmission lines that incorporates super-cooled gases with specific ceramics or metals in the lines resulting in more efficient operations than traditional copper wires. Superconductivity virtually eliminates electrical resistance for improved capacity and more efficient power interconnections.

Humvee House
Outsized homes favored by some Americans that require more energy to heat and illuminate and usually contain more devices that draw electricity than normally sized dwellings. Recent market data indicates that these energy hogs are losing appeal. Also known as a Hummer house.

HVAC (Heating, Ventilation, Air Conditioning)
The system or systems that condition air in a building. According to the DOE (Department of Energy), these residential and commercial building systems use 16 percent of all energy in the US.

HVAC conservation feature
A building feature designed to reduce the amount of energy consumed by the heating, cooling, and ventilating equipment.

HVDC (High Voltage Direct Current)
Direct current that allows controlled transmission of large amounts of power efficiently over very long distances in narrower rights-of-way.

HVRT (High Voltage Ride Through)
An event that is usually caused by a sudden increase in power from an intermittent renewable energy source like wind or solar that is integrated to the electrical grid and could cause grid instability if not mitigated with controlling actions by a grid operator.

Hybrid transmission line
A double-circuit line that has one alternating current and one direct circuit. The AC circuit usually serves local loads along the line.

Hydraulic head
The pressure of water based on elevation, usually expressed in feet of head or in pounds per square inch. Also known as hydrostatic head.

Hydroelectric plant
An electrical generation facility using falling water to spin a turbine generator. Hydroelectric power supplies approximately twenty-four percent of the world's electricity and about twelve percent of electricity in the USA.

Hydrology
The scientific study of the cycle from precipitation to re-evaporation or return of water to a water source. It is used to predict rates and amounts of runoff, estimate spillway and reservoir capacities, and estimate water supplies for water resource management.

Hydrometry
The measurement of water flow variables for water resources management.

Hydrothermal
A form of geothermal energy using natural underground reservoirs of hot water or steam for direct use or generation of electricity.

I

I2G (Industry to Grid)
An acronym that describes the connection of industrial buildings and facilities to the Smart Grid. It is also a working group within NIST (National Institute of Standards and Technology) focused on this subject. (www.nist.gov/smartgrid/)

I3P (Institute for Information Infrastructure Protection)
A US-based consortium of leading universities, national laboratories, and nonprofit institutions dedicated to improving R&D to protect the national information infrastructure against catastrophic failures. The institute's main role is to coordinate a national cyber security R&D program and help build bridges between academia, industry, and government. (http://www.thei3p.org/)

IAEE (International Association of Energy Economics)
A global, nonprofit, professional organization with members in over 70 nations. It provides a forum for ideas, experiences, and issues among professionals interested in the field of energy economics, along with research and educational development for members. (http://www.iaee.org/en/)

IAPP (International Association of Privacy Professionals)
A not-for-profit membership association of privacy resources with a mission to define, support and improve the global privacy profession through networking, education and certification. (https://www.privacyassociation.org/)

IAREC (Inter-regional Association of the Regional Energy Commissions)
A voluntary, non-commercial union established by the Russian Federation to develop principles, rules, norms and standards of economic regulation in the energy sector. It also focuses on development and implementation of uniform tariff policies at federal and regional levels. (http://www.mtu-net.ru/marek/eng/index_marec_e.html)

IBC (International Building Code)
A building model code developed by the International Code Council (ICC). For specific building functions, IBC seismic codes can require that systems critical to life safety and fulfilling the building's intended purpose, such as local power generators, remain online immediately after a seismic event. This code has implications for backup and emergency power systems, requiring the system be certified compliant to the same seismic category designation as the building.

IBEW (International Brotherhood of Electrical Workers)
One of the largest unions of utility trade employees, the IBEW represents linemen, groundmen, electricians, control room operators, meter installers, meter readers, field reps, clerical workers, cable splicers, substation electricians, and communications techs, and other job roles. The IBEW also represents employees in gas and telecom utilities. (http://www.ibew.org/)

ICAP (IEEE Conformity Assessment Program)
A branch of IEEE's Industry Standards and Technology Organization, which provides programs and other support for industry to bridge the development of IEEE standards with conformity assessments. The purpose of this is to accelerate widespread acceptance and the enabling of new products and the advancement of technology.

ICAP (Installed Capacity)
A monthly market run by an ISO (Independent System Operator) that provides generators compensation for locating units in specific regions based on the net capacity the unit provides to the market after accounting for forced outages at the unit.

ICC (Insulated Conductors Committee)
A technical committee for the IEEE, involved in writing standards and guidelines for use in installation and test of underground transmission and distribution cables.

ICC (International Code Council)
A nonprofit association dedicated to developing a set of comprehensive and coordinated national building safety and fire prevention construction codes for residential and commercial buildings. It includes seismic codes that impact siting and certification of power generation equipment.

ICCP-TASE.2 (Inter-control Center Communications Protocol - Telecontrol Application Service Element.2)
This standard, also known as IEC 60870-6, enables the exchange of time-critical information between utility control systems like SCADA (Supervisory Control And Data Acquisition) and power plant dispatching applications via WAN and LAN. This standard has applications with IEDs (Intelligent Electronic Devices).

ICS (Industrial Control Systems)
Hardware, firmware, and software and networks that provide locally and/or remotely displayed data and control capabilities to manage systems ranging from transmission and distribution substations, generation plants, water and wastewater, and any other industrial processing operations. SCADA (Supervisory Control and Data Acquisition) systems are examples of ICS.

ICT (Information and Communications Technology)
Computing and communicating technologies covering hardware and software that create, communicate, store, and manage information. ICT includes fixed and mobile devices and the software applications and services associated with them.

IDC (Interchange Distribution Calculator)
A mechanism used in the Eastern Interconnection to calculate the distribution of interchange transactions and includes a database of all interchange transactions and a matrix of the distribution factors for the Eastern Interconnection.

IDC (Interest During Construction)
An accounting term also known as Allowance for Funds used During Construction.

IDEA (International District Energy Association)
A global, nonprofit trade association that facilitates the exchange of information among district energy professionals operating district energy systems owned by utilities, municipalities, hospitals, military bases and airports around the world. It promotes energy efficiency and environmental quality through the advancement of district heating, district cooling and cogeneration. (http://www.districtenergy.org/)

IDIS Association (Interoperable Device Interface Specifications)
A nonprofit association established to maintain and promote publicly available technical interoperability specifications based on open standards and supports their implementation in interoperable products. IDIS is an association for smart metering companies which are committed to providing interoperable products based on open standards. The current members are Iskraemeco, Itron and Landis+Gyr. (www.idis-association.com)

Idle capacity
The component of operable capacity to generate electricity that is not in operation and not under active repair but can be placed in operation within 30 days, and capacity to generate electricity not in operation but under active repair that can be completed within 90 days.

IDS (Intrusion Detection System)
A security methodology that monitors unauthorized entry or unauthorized activity in sensitive systems. It may also act to block selected users or source IP addresses from accessing the network. It is one of several layers of security recommended for utility networks.

IDSM (Integrated Demand-Side Management)
A recent concept that proposes integrating energy efficiency and demand response programs along with distributed generation to leverage synergies for both end users and utilities. It includes a foundation built on smart meters and dynamic pricing. The benefits would be lower energy costs, increased grid reliability, and better returns on DSM investments by utilities and end users.

IEA (International Energy Agency)
An intergovernmental organization that acts in a policy advisory role to 28 member countries for balanced energy policy making, energy security, economic development, and environmental protection. Current work focuses on climate change policies, market reform, energy technology collaboration, and outreach to the rest of the world, especially major consumers and producers of energy like China, India, Russia, and the OPEC countries. It will provide an assessment of current trends and prospects to 2020 of transmission investments in IEA countries and in China, including major domestic and cross-border projects in North America and Europe, including discussion of Smart Grid opportunities and challenges. (http://www.iea.org/index.asp)

IEC (International Engineering Consortium)
A nonprofit organization sponsored by universities and engineering societies dedicated to continuing education for the US electronics industry. It also provides research, publications, and service programs for the international information industry. (www.iec.org)

IEC (International Electrotechnical Commission)
A not-for-profit, global organization that prepares and publishes international standards for all electrical, electronic, and related technologies. These standards are the core of the World Trade Organization's Agreement on Technical Barriers to Trade. The IEC also manages three global conformity assessment systems to certify if

equipment conforms to its standards. It works with the ISO (International Organization for Standardization), ITU (International Telecommunications Union), and IEEE (Institute of Electrical and Electronics Engineers). Members include manufacturers, providers, distributors, vendors, consumers, users, governmental agencies, professional societies, trade associations, and standards developers from national standards bodies. For North America, the members are the Standards Council of Canada, ANSI (American National Standards Institute), and NEMA (National Electrical Manufacturers Association). (http://www.iec.ch/)

IEC 60255-22-x
This set of standards specifies requirements for measuring relays and protection equipment for power systems, including the control, monitoring, and process interface equipment. Each part includes a different type of electrical disturbance test to ensure equipment integrity under electrical surge or loss.

IEC 60870-2-1
This standard covers the operating conditions of power supply and electromagnetic equipment. It includes documentation, safety, compatibility, and fire hazard testing. Its main function is in telecontrol and telemetering SCADA applications.

IEC 60870-5
This standard developed by IEC TC 57 WG 3 is used for the telecontrol of SCADA and power system automation applications. It includes data link transmission services, coding of information elements, and transmission frame formats between central telecontrol stations and outstations with direct data circuits.

IEC 60870-6
A standard, also known as TASE.2 (Telecontrol Application Service Element.2) or ICCP (inter-control Center Communications Protocol), used

internationally for WAN and LAN communications between utility control centers and with SCADA (Supervisory Control And Data Acquisition) systems and other engineering systems found in control centers.

IEC 60929
This standard establishes performance requirements for electronic ballasts up to 1,000V at 50-60 Hz. These ballasts are used to operate and control AC high power fluorescent lighting installations.

IEC 61131-3
A standard that established a vendor-independent programming language for industrial automation and PLCs (programmable logic controllers). It improves the time to deploy and reduces programming errors.

IEC 61158 (FieldBus)
This is a collection of industrial computer network protocols used for real-time control of distributed assets. This includes automation via an internet connection and PLCs (Programmable logic controllers). Fieldbus is the communication standard that links remotely operated small scale assets such as sensors, actuators or valves.

IEC 61326
This set of standards specifies the requirements for electromagnetic emissions and immunity for electrical equipment operating with a battery less than 1,500 Volts. Included are testing and measurement procedures, computing equipment, and PLCs (Programmable Logic Controllers) intended for industrial and non-industrial locations. Part 2-2 focuses on the distribution grid, while part 2-6 focuses on medical equipment.

IEC 61334
A series of standards that specify management of data communications protocols and systems for low speed PLC (Power Line Communications) in the distribution grid.

IEC 61499
This standard is the basis from which distributed control and automation systems are developed. It defines an open architecture that gives a generic model for the creation of distributed control applications.

IEC 61508
This standard defines functional safety requirements for electrical, oil and gas, electronic, and programmable electronic safety systems. It is intended to provide assurance that safety related systems will offer the expected risk reductions for safe operations.

IEC 61511/ ISA 84
Similar in function to IEC 61508, this standard sets requirements and selection criteria for components and instrumented systems used in safety functions within the process industry sector.

IEC 61588
The precision clock synchronization protocol for networked measurement and control systems standard enables system-wide synchronization accuracy in the sub-microsecond range with minimal network and local clock computing resources. It is used in measurement and control systems like PMU (Phasor Measurement Unit) deployments.

IEC 61850
A standard used for field communications, such as substation automation, distributed generation (using renewables and fuel cells), SCADA (Supervisory Control And Data Acquisition) communications, and distribution automation. The standard is considering inclusion of PHEV (Plugin Hybrid Electric Vehicles). It is also a candidate for revision and expansion to include more distributed generation functions, phasor technology, and communications. It currently consists of 10 parts with parts 6 through 10 considered the primary parts:
Part 1 – Basic Principles

Part 2 – Glossary
Part 3 – General Requirements
Part 4 – System and Project Management
Part 5 – Communication Requirements
Part 6 – Substation Automation System Configuration
Part 7 – Basic Communication Structure
Part 8 – Mapping to MMS and Ethernet
Part 9 – Sampled Measured Values and Mapping to Ethernet
Part 10 – Conformance Testing

IEC 61850ug (IEC 61850 Technical Subcommittee)
A group that is part of UCAIug and administers the identification and resolution of technical problems for the IEC 61850, focused on systems and network management. The group's work focuses on system design, monitoring and management tools for substations, and the backup of the substation LAN hardware, software, communications, and management functions. This project is also sponsored by American Electric Power and EPRI. (www.IEC61850ug.org)

IEC 61850-7-420
This is the communications standard for DER (distributed energy resources) such as generation devices and energy storage and microgrids. The goal of this standard is to create a single international standard for the communication as well as control interface of DER devices.

IEC 61850-90-4
This standard focuses on the development of a local area network to assist in substation automation. It also outlines potential improvements and upgrades to existing networking and substation automation equipment.

IEC 61850-90-5
A proposed amendment to IEC61850 that describes how to use this standard to transmit synchrophasor information according to IEEE C37.118.

IEC 61850-90-11
This standard is for communication networks and systems for power utility automation, with a focus on methodologies for logic modelling within IEC 61850 based applications.

IEC 61850-100-1
This standard in developmental phase addresses the commissioning of tests for IEC 61850 based systems.

IEC 61851
This standard outlines the requirements for charging equipment for electric road vehicles rated up to 1,500 Volts. Included are required characteristics and operating conditions of the electric supply device. Important facets of this standard are sections on operator electrical safety, as well as communication between the charging station and the electric vehicle.

IEC 61968
A standard used to define the CIM (Common Information Model) for distribution management and AMI (Advanced Metering Infrastructure) back-office interfaces.

IEC 61968-14-1
A harmonization effort to map MultiSpeak Version 4.0 and specific subparts of the IEC 61968 standard so that an electronic translation occurs between MultiSpeak service calls and CIM (Common Information Model) messages. This standard is planned for inclusion in the existing IEC 61968 standards family.

IEC 61968-14-2
A standards development activity to create CIM (Common Information Model) profiles with the same functionality as corresponding MultiSpeak Version 4.0 interface definitions. This is planned for inclusion into the existing IEC 61968 standards family.

IEC 61968/61970-5-101,103,104

This is a standard used for the Transmission CIM (Common Information Model). These subsections address CIS (Component Interface Specification), common services, generic data access and HSDA (High Speed Data Access) respectively.

IEC 61970

A standard used for the Transmission CIM (Common Information Model).

IEC 62056

This standard outlines hardware and communications protocol requirements for the data exchange, meter reading, and tariff and load control of smart meters. This standard is the international version of the DLMS (Device Language Message Specification) protocol.

IEC 62061

This standard specifies requirements for the design, integration, and validation of functional safety-related electrical, electronic, and programmable electronic control systems.

IEC 62351

A security standard that impacts the other IEC standards. It is focused on IEC protocols, network and system management, and role-based access control. Parts 1 through 8 address information security for power systems and control operations.

IEC 62351-1:8

A security standard developed by TC 57 WG 15 that impacts the other IEC standards. It is focused on the security of IEC protocols, network and system management, and role based access control. Parts 1 through 8 address information security for power systems management and control operations.

IEC 62400-25-1

This standard outlines the structuring principles and documentation for technical products within the power generation, distribution, and electrical engineering space. These devices are organized numerically under the standard by purpose and task.

IEC 62443

A series of 14 standards that address network and system security for industrial automation and control systems. Four sub series address general, manufacturer, system integrator, and owner-operator perspectives.

IEC 62559

A PAS (Publicly Available Specification) that describes a methodology for power system engineers to identify and document user requirements for automation systems, based on their utility business needs. It was originally developed as part of the IntelliGrid Architecture from EPRI (Electrical Power Research Institute) as a means to implement the "IntelliGrid vision."

IEC 62591 Ed.1.0

A standard for wireless communication in industrial communications for process industries. It includes physical layer definition and protocol specifications, data-link and application layer services and protocols, security, and gateway procedures. It supersedes IEC/PAS 62591 published in 2009.

IEC CISPR (Comité International Spécial des Perturbations Radioélectriques) The International special committee on radio interference, created by the IEC. The primary focus of this special committee is to create electromagnetic compatibility standards to protect radio reception in the range of 9 kHz to 400GHz from interference. Common interference sources are electricity supply and transport systems, all types of electrical appliances, and ignition systems.

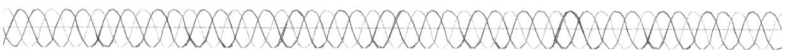

IEC CISPR 22
This is the international standard of measurement for spurious signal and radio disturbance emitted by IT equipment. It has been published as IEEE C63.022 as an American standard. See IEEE c63.022 for additional information.

IEC CISPR 24
This standard is based on CISPR 22 (IEEE C63.022), and defines requirements to establish a level of immunity to environmental factors for electrical equipment. The tests the equipment must pass are to prevent electrostatic discharges as well as radio disruptions above a certain level.

IEC PC (Project Committee) 118
This project committee outlines information exchange protocols and interface requirements for demand response equipment. It specifically connects demand-side equipment and systems into the Smart Grid. Also in development is a standardized smart grid user interface.

IEC Subcommittee 17C
This is the subcommittee dedicated to preparing international standards covering prefabricated assemblies. These standards include communication requirements of high voltage switchgear assemblies, measuring and regulating equipment, and gas insulated transmission lines.

IEC Subcommittee 22F
This is the subcommittee responsible for the standardization of electronic power conversion and semiconductor switching systems. These standards are applied to electrical transmission and distribution systems including control, protection, and monitoring of high voltage direct converters, controlled series capacitors, and other applications utilizing semiconductors.

IEC Subcommittees 86A, B, C
This group of subcommittees focuses on developing standards for fiber optic cables, systems, and networks.

IEC Technical Committee 3
This is the technical committee dedicated to information structures, documentation, and graphical symbols. Its scope is methods and rules associated with the human interpretation of information, as well as methods and rules associated with processing information in a computer sensible form. TC 3 also develops the standards for graphical symbols used in diagrams and on equipment.

IEC Technical Committee 8
This is the technical committee dedicated to developing standards for the entirety of the electricity supply chain from production to end use. These areas include electrical system reliability and connections, network and communications safety and security, metering and data exchange, and charging mechanisms.

IEC Technical Committee 13
This is the technical committee dedicated to the standardization of AC and DC electrical measurement and control, including modern smart metering equipment. These materials include the standards of transmission, distribution, and supply networks as part of the scope of their energy management and Smart Grid activities.

IEC Technical Committee 21
This is the technical committee responsible for the creation of standards pertaining to secondary cells and batteries and their safety regulations. These include photovoltaic electricity storage systems, batteries used for the propulsion of electric and hybrid road vehicles, and RFID tags.

IEC Technical Committee 23
This is the technical committee that oversees the development of standards for electrical accessories in household, commercial, and industrial applications. The accessories are fixed infrastructure for use with appliances or other electronic components and devices. These components include conduit systems, switches, plugs and socket outlets, cable reels, and adaptors.

IEC Technical Committee 32
This is the technical committee that is responsible for the development of standards regarding the specifications for all types of fuses. These standards include direction for installation and operation of fuses, as well as requirements for compliance test methodology.

IEC Technical Committee 38
This is the technical committee that develops the standards for instrument transformers and their internal components. The standards govern current, inductive, and capacitive voltage transformers, as well as their internal sensing and data conversion interfacing devices.

IEC Technical Committee 57
The committee that develops and maintains international standards for power systems control equipment and data systems, including EMS (Energy Management System), SCADA (Supervisory Control And Data Acquisition), distribution automation, teleprotection, and associated information exchange for real-time and non-real-time information used in the planning, operation, and maintenance of power systems. It is also referred to as the power systems management and associated information exchange committee. (http://tc57.iec.ch/index-tc57.html)

IEC Technical Committee 57 AHG (Ad-Hoc Group) 8
This Ad-Hoc Group reviews and recommends the use of IPv6 as a network protocol for IEC TC 57 standards. The review includes cyber-security issues,

management requirements, and impact studies on other IEC standards. Uses for the product of this group are utility communication on public and private networks, substation communication, and control center interconnection, as well as other Smart Grid communications that use an Internet connection.

IEC Technical Committee 57 Working Group 3
This working group focuses on the development of telecontrol protocols. The standards created by WG 3 are implemented in IEC 60870-5, which defines requirements for SCADA and power system automation applications.

IEC Technical Committee 57 Working Group 9
This working group is responsible for the development of standards pertaining to grid automation using PLC (Power Line Carrier) systems.

IEC Technical Committee 57 Working Group 10
This working group is tasked with the development of communication standards for substations, which includes functional architecture, power system IED communication, and associated data models.

IEC Technical Committee 57 Working Group 13
This working group produces standard interface specifications or APIs for an electric utility power control center or EMS (Energy Management System). The API specifications include that the software solution must be easily installable with no modification of source code across a variety of platforms.

IEC Technical Committee 57 Working Group 14
This working group is establishing requirements for a standardized interface for DMS (Distribution Management Systems). The standard will define interfaces for the major components of DMS.

IEC Technical Committee 57 Working Group 15
This working group sets security standards for power system information technology infrastructure, as well as M2M communications. The impacts communications protocols for IEC 60870, 61850, 61970, and 61968. Its focus is the improvement of power system reliability through system communication security

IEC Technical Committee 57 Working Group 16
This working group is developing standards for deregulated electricity market communications including market participants and operators. Standards developed by this group adhere to TC 57's CIM (Common Information Model) using the XML information exchange format.

IEC Technical Committee 57 Working Group 17
This working group of Technical Committee 57 is responsible for the creation of standards and guidelines for communications systems interconnecting DER (Distributed Energy Resources).

IEC Technical Committee 57 Working Group 19
This working group focuses on maintaining long term interoperability between the standards developed by Technical Committee 57. A primary objective of WG 19 is to ensure standard integration into IEC's CIM (Common Information Model) and SCL (Substation Configuration Language).

IEC Technical Committee 57 Working Group 21
This working group identifies use cases involving systems connected to the electrical grid. The standards developed from the use cases focus on requirements for system interfaces and communication protocol profiles for grid-connected systems, such as building automation and DER connected via wireless and wired communications.

IEC Technical Committee 64

This is the technical committee that defines the standards for the safety of electrical installations against electric shock. This includes installation requirements and basic safety protocols and requirements.

IEC Technical Committee 65

This is the technical committee responsible for the development of standards for systems and elements used in industrial process measurement, control, and automation. Included in these standards are safety and security protocols, as well as standards for electrical, hydraulic, and mechanical control systems.

IEC Technical Committee 69

This is the technical committee dedicated to preparing international standards for road vehicles, specifically those vehicles totally or partially electrically propelled from self-contained power sources. This includes electric industrial trucks.

IEC Technical Committee 77

This is the technical committee dedicated to developing standards that protect equipment, systems, and installations from the unintentional generation, propagation, and reception of low to high frequency electromagnetic energy. This is known as electromagnetic compatibility (EMC).

IEC Technical Committee 82

This is the technical committee responsible for the development of international standards for photovoltaic (PV) energy systems. This includes elements within the entire PV system, as well as interfacing with the electrical system to which energy is supplied.

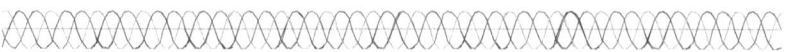

IEC Technical Committee 88
The objective of this subcommittee is to develop international standards for wind turbines that convert wind energy into electrical energy. These standards cover the internal mechanical and electrical systems of the turbines, design requirements, and testing methodologies. The goal of the standards is to provide a base for design, quality assurance and integrity certification.

IEC Technical Committee 95
This is the technical committee devoted to standardizing measuring relays and protection equipment. These systems are primarily used by electricity supply utilities and industrial users.

IEC Technical Committee 105
This is the technical committee dedicated to developing international standards for fuel cell technologies. These standards include stationary fuel cells, auxiliary power units, and mobile fuel cells used for vehicle propulsion.

IEC Technical Committee 114
This is the technical committee dedicated to preparing international standards for marine energy conversion systems. These systems are wave, tidal, and current energies being converted into electricity. The standards cover environmental impact evaluation, design and safety requirements, and factory testing methods.

IEC Technical Report 61000
This report provides guidance for EMC (electromagnetic compatibility) requirements for the limitation of power frequency conducted harmonic current emissions from electrical equipment. It includes environmental classifications, electromagnetic emission limits, as well as testing and measurement techniques. This report influences the IEC 61000 series of standards.

IEC Technical Report 62357-1
This report concerns power systems management and its associated information exchange, with a focus on reference architecture for information exchange. It contains an overview of standards that provide guidelines for application to distribution, transmission, and generation systems.

IECC (International Energy Conservation Code®)
A set of building codes that promotes energy conservation through efficiency in envelope design, mechanical systems, lighting systems, and the use of new materials and techniques. The codes are updated every three years by the ICC (International Code Council) and were updated for 2009.

IED (Intelligent Electronic Device)
A class of devices that provides protection, control, and/or monitoring capabilities for power system equipment. An IED includes microprocessors to control power system equipment, such as circuit breakers, transformers, and capacitor banks, and communications capabilities for remote operations. IEDs include fault locators, circuit breaker recorders, digital protective relays, PMUs, and RTUs.

IEE (Institute for Electric Efficiency)
A program of the Edison Foundation with the objectives to advance energy efficiency practices and DR among electric utilities, share information and experiences about efficiency program deployments, and create a collection of effective business models, practices, and processes. (http://www.edisonfoundation.net/iee/)

IEEE (Institute of Electrical and Electronics Engineers)
A global, nonprofit professional organization for the advancement of technology in a range of areas, including electric power, computers and telecommunications, and consumer electronics, among others. It is composed of a number of societies, of which the Power and Energy Society

is one example, targeted to electric power professionals. Members rely on IEEE as a source of technical and professional information, resources, and services. (http://www.ieee.org/portal/site)

IEEE 80
The guidance for safe grounding practices in AC substation design, primarily outdoors, but can pertain to indoor designs, too. It includes testing procedures focused on earth resistivity ground impedance and earth surface potentials of a ground system (81) and testing procedures covering impedance and safety characteristics of large extended or interconnected grounding systems (81.2). 81 and 81.2 are being combined into one standard by 2011.

IEEE 400 ™
This is a series of four standards for test methods used with shielded power cable systems rated 5 kV and above.

IEEE 473
Recommended practices for interior and exterior electromagnetic site surveys, which measures electric and magnetic fields and interference between the frequency range of 10kHz and 10GHz. It delivers a standardized means of obtaining knowledge about the electromagnetic environment any given device will be operating in, with forewarning of potential sources of interference. It is currently withdrawn and in the process of updating via Technical Committee 3.

IEEE 519
A standard that describes practices and recommendations for harmonic control in electrical power systems to limit voltage distortions. This standard has impacts on energy storage at the point of common coupling between the equipment and the utility's electrical grid.

IEEE 802
The collection of LAN and MAN standards. The most widely used standards include Ethernet family, Token Ring, Wireless LAN, Wireless PAN, Wireless MAN, Bridging, and Virtual Bridged LANs.

IEEE 802.1AB
A standard that defines the protocols for device discovery on LANs or MANs and specifies network management information to enable multi-vendor interoperability.

IEEE 802.2
A standard that specifies the Logic Link Control protocol and is used in Ethernet and Token Ring LANs to manage data communications.

IEEE 802.3
The Ethernet standard covering CSMA/CD (Carrier Sense Multiple Access with Collision Detection) access method and the physical layer specification.

IEEE 802.11
The wireless LAN family of standards. It includes the WiFi protocol. 802.11s concerns standardizing the hardware of wireless mesh networks. 802.11i is also known as WPA2, or RSN (Robust Security Network) and uses the Advanced Encryption Standard (AES) block cipher to provide secure communications. 802.11a/b/g/n are all updates to the Wi-Fi standard, which incrementally increase performance, bandwidth, and power consumption respectively. The most recent update is 802.11AC, superseding 802.11n.

IEEE 802.11AC
The successor to 802.11n, this emerging standard operates on the same technology currently in use for "n", but greatly improves the 5.0Ghz band,

as well as the scalability of WiFi networks. It achieves higher data rates and link reliability.

IEEE 802.15
The wireless PAN family of standards. Bluetooth is based on 802.15.1, and ZigBee and 6LoWPAN are based on 802.15.4. This latter standard is the focus of activity to create a global standard to facilitate utility telemetry networks for large numbers of meters.

IEEE 802.16
The WiMAX family of standards. WiMAX products accommodate fixed (802.16d) and mobile (802.16e) applications. 802.16m is the next-generation standard beyond 802.16e-2005, and it is considered to be a strong candidate as a 4G technology.

IEEE 802.16.1™-2012
IEEE Standard for WirelessMAN-Advanced Air Interface for Broadband Wireless Access Systems. The standard provides an enhanced air interface and improved capacity for metropolitan-area networks that utilities can use for smart grid machine-to-machine communications, as well as mobile voice-based applications, with support for Ethernet as well as IP interfaces. IEEE 802.16.1-2012 is a new standalone version of the technology first specified in IEEE 802.16m™-2011 and designated by the ITU as IMT-Advanced. Further enhancement relevant to smart grid applications are provided in IEEE 802.16.1b™-2012, an amendment providing "Enhancements to Support Machine-to-Machine Applications."

IEEE 802.20
The mobile broadband wireless access systems standard for vehicles with a wide range of licensed spectrum. 802.20a covers PHY and MAC layers, and 802.20b covers virtual bridging through service interface mapping with 802.20 MAC.

IEEE 802.24 Technical Advisory Group
This group acts as a liaison with other regulatory and standards developing agencies, government agencies, and industry organizations regarding the 802 standards series. It answers technical questions regarding the use of 802 standards in Smart Grid applications.

IEEE 1020
A standard that addresses small hydro monitoring and control requirements.

IEEE 1031^{TM}
Standard that provides the functional specifications for transmission static VAR compensators.

IEEE 1127
A description of important community acceptance and environmental compatibility considerations during the planning, design, construction, and operation of substations. On-site generation and telecom facilities are not included.

IEEE 1159.3
A recommended file format for the transfer of power quality data and other data. It defines a highly compressed storage scheme suitable for the transfer of power quality measurement and simulation data for all equipment vendors.

IEEE 1250^{TM}
Standard that describes identification and improvement of voltage quality in power systems.

IEEE 1344
A protocol for real-time synchrophasor data, replaced by IEEE C37.118.

IEEE 1366-2012
Draft standard that addresses reliability index definitions and factors that impact their calculations that are used for distribution systems, substations, circuits, and defined regions.

IEEE 1377
A standard that provides the application layer format for the utility data to be passed between an End Device and a computer. It provides a plug and play environment for field metering devices. It is co-published as ANSI C12.19.

IEEE 1379
Guidelines for communications and interoperability between IEDs (Intelligent Electronic Devices) and RTUs (Remote Terminal Units) in substations with a limited subset of two existing communications protocols.

IEEE 1402
Guidelines that identify and discuss security issues related to human intervention during the construction, operation (except for natural disasters), and maintenance of substations.

IEEE 1451
The standards family that covers a common set of data and control formats for sensors and actuators and is part of the standards composition for the Internet of Things. It is the same standard as ISO/IEC 21451-7.

IEEE 1491
This standard is the guide for the selection and use of stationary battery monitoring equipment applications. These applications define what parameters must be observed by battery monitoring equipment.

IEEE 1547

The standards series that outlines criteria and requirements for distributed energy resources' physical and electrical interconnection to the power grid. It provides requirements relating to performance, operation, testing, safety considerations, and maintenance of the interconnection. Some standards are approved, others are in the review process.

IEEE 1547.1

Provides conformance test procedures for equipment interconnecting distributed energy resources with the electrical grid.

IEEE 1547.2

The application guide for interconnecting distributed energy resources with the grid.

IEEE 1547.3

The guide for monitoring, information exchange, and control of distributed energy resources.

IEEE 1547.4

The guidelines for the design, operation, and integration of distributed energy resources that function as island systems (microgrids) with the electrical grid.

IEEE 1547.8

An expansion of the 2008 standard for interconnecting distributed resources with power systems. 1547.8 is the current draft project for establishing methods and procedures that provide supplemental support for implementation strategies of Standard 1547.

IEEE 1560

This is a standard for methods of measuring RF (radio frequency) interference in power line in the ranges of 100 Hz to 10 GHz. It is

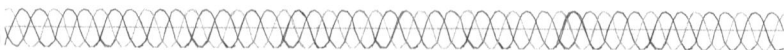

specifically to govern products designed to mitigate interference conducted on power lines.

IEEE 1588™
A standard that defines a protocol enabling clock synchronization in networked measurement and control systems.

IEEE 1591.1
A standard covering manufacturing, testing, and procurement specifications for use with OPGW (OPtical Ground Wire) hardware used in transmission and distribution grids for both electricity and communications signaling.

IEEE 1613-2009
This is the standard for environmental and testing requirements for communications networking devices in electric power substations.

IEEE 1615
Guidelines for communication and interoperation of substation devices connected on an IP network.

IEEE 1646
A standard that defines communication delivery times for information exchanged internally and externally to substation integrated protection, control, and data acquisition systems.

IEEE 1675
A standard that defines testing and verification for common hardware like couplers and enclosures for BPL (Broadband Powerline) installations and provides suggested installation methods to ensure compliance with codes.

IEEE 1686
The draft standard for IED's (intelligent electronic devices) cyber security capabilities. The standard defines the functions and features to be provided in IEDs to accommodate critical infrastructure programs (CIP). It covers access, operation, configuration, firmware revision, IED data retrieval, and encryption of IED communications.

IEEE 1701
This standard defines communications criteria with meters using optical ports. Other devices include hand-held readers, portable computers, or other electronic communications devices. It provides the optical port lower layers communication protocol for meters. It is co-published as ANSI C12.18.

IEEE 1702
This standard details the criteria required for communications between utility end devices and utility hosts using modems connected to the switched telephone network. The protocol specified in this standard is designed to transport data in the table format provided in IEEE 1377 or ANSI C12.19 standards. It is co-published as ANSI C12.21.

IEEE 1703
This standard defines the means to transport the utility end device data tables via LANs or WANs in AMI deployments. It provides the lower layers communication protocol. It is co-published as ANSI C12.22.

IEEE 1775
A standard for PLC (Power Line Carrier) equipment and its EMC (Electromagnetic Compatibility) requirements, as well as testing and measurement methods. Included in the standard are EMC criteria for BPL (Broadband over Power Line) communications and installations.

IEEE 1808[TM]
The standard for collecting and management transmission line inspection and maintenance data.

IEEE 1815[TM]
The standard for DNP3 (Distributed Network Protocol 3) that is used for communications between devices to achieve interoperability. It's Secure Authentication (SA) protocols are being enhanced to add security to industrial processes.

IEEE 1815.1
This is an active project under development in a working group, the goal being mapping between the currently established IEEE 1815 and IEC 61850. The two priorities are mapping between IEEE 1815 based masters and IEC 61850 based servers, and mapping between IEC 61850 based clients and IEEE 1815 based outstations.

IEEE 1888
Standard that describes remote control architectures for communities, groups of intelligent buildings or campuses, and metropolitan networks to enable interconnection, collaborative services, remote monitoring and control. It impacts the Internet of Things.

IEEE 1901
A standard that addresses MAC (Medium Access Control) and PHY (Physical) Layer specifications for BPL (Broadband Powerline) networks. The standard uses transmission frequencies less than 100 MHz and applies to all classes of BPL devices, including devices used for the first-mile/last-mile connection to broadband services and BPL devices used in buildings for LANs, Smart Energy applications, transportation platforms (vehicle) applications, and other data distribution over AC power lines. It covers mechanisms for interoperability and security issues.

IEEE 2030™

This standard provides guidelines to understand and define Smart Grid interoperability of energy, communications, and information technologies with the electric power system and end-use applications and loads. This standard addresses two-way power flow with communication and control. It covers demand response, WASA (Wide Area Situational Awareness), storage, distribution, metering infrastructure, security, and management of devices like sensors. It also discusses best practices for the Smart Grid. (http://grouper.ieee.org/groups/scc21/2030/2030_index.html)

IEEE 2030.5

This standard incorporates Smart Energy Profile (SEP) 2.0. It defines application message exchange mechanisms, the exact messages exchanged, and the security features used to protect the application messages. This enables utility management of the end user energy environment, including demand response, load control, and time of day pricing among other functions.

IEEE 3000 Standards Collection™

A series of standards that set guidelines for the industrial and commercial power generation processes.

IEEE 3001.8

This standard for the instrumentation and metering of industrial and commercial power systems addresses recommended metering practices to achieve more efficient energy management, as well as issues that must be taken into account when applying the most recent metering technologies.

IEEE 3004.1

This standard for the application of instrument transformers in industrial and commercial power systems primarily concerns the recommended practice for selection of instrument transformers.

IEEE 3006.7
This standard determines the reliability of 7x24 continuous power systems in industrial and commercial facilities. It deals with the recommended methods of testing the reliability of the systems, evaluation of test results, and implementing changes based on those results to maximize system reliability.

IEEE 3006.9
This standard is for collecting data for use in evaluating the reliability, availability, and maintainability of industrial and commercial power systems. It is designed to be a recommended practice aid to engineers working on the electrical design of power systems.

IEEE 3007.1™
This standard for industrial and commercial power plant maintenance, operations, and safety covers documentation, system operation, and control responsibilities.

IEEE 3007.2 ™
This standard for industrial and commercial power plants covers equipment maintenance and testing.

IEEE 3007.3™
This standard for industrial and commercial power plants covers electricity hazards and safety.

IEEE C2-2007
This IEEE and ANSI standard is also known as the National Electric Safety Code. It addresses human safety during the installation, operation, and maintenance of electric supply and communications lines.

IEEE C37.1
A standard for SCADA (Supervisory Control And Data Acquisition) and automation systems that provides the definition, specification, performance analysis, and application of these systems in staffed and unattended substations with some automatic control functions.

IEEE C37.2
The Electrical Power System Device Function Numbers, Acronyms, and Contact Designations standard that defines the acronyms and function numbers for use with devices and processes in substations, generating plants, and sites with power utilization and conversion equipment.

IEEE C37.90.1
A standard identifying surge tests for protective relays and relay systems, including test procedures, acceptance criteria, and documentation requirements. It has implications for energy storage.

IEEE C37.90.2
A standard that defines test for relays and relay equipment to determine their immunity to electromagnetic interference caused by mobile radio and wireless communications devices. It has impacts for energy storage assets.

IEEE C37.91
Describes guidelines for protecting 3-phase power transformers used in transmission and distribution systems along with discussion about technical problems with protection systems.

IEEE C37.92
A standard that defines interface connectivity between voltage or current transducer systems and electronic relays or other substation measuring equipment.

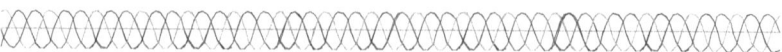

IEEE C37.93
Guide with information and recommendations for applying audio tones over voice grade channels for power system relaying, including telecom equipment and lines, application principles, installation, and testing.

IEEE C37.94
A standard that defines optical-fiber interconnection of different vendors' relays with different vendors' multiplex equipment, without any restriction on the content of the N times 64-kilobit-per-second data, using up to 2 kilometers of 50-micrometer or 62.5-micrometer multimode fiber. Requirements for communications timing are also included.

IEEE C37.95
Guide with information on different protective relaying practices for the utility-consumer interconnection. It addresses applications with service to a consumer that normally requires a transformation between the utility's supply voltage and the consumer's voltage requirement.

IEEE C37.101
Guide to provide generator ground protection.

IEEE C37.102
Guide for synchronous generator protection, typically large hydraulic, steam, and combustion turbine generators.

IEEE C37.104
A guide that identifies practices for automatic reclosing of AC distribution and transmission lines.

IEEE C37.106
Guide that describes use of relays to protect generating plant equipment from damage caused by abnormal frequencies, with a focus on fossil fuel, nuclear, and combustion turbine installations.

IEEE C37.111
The standard Common Format for Transient Data Exchange for power systems or COMTRADE that defines a file format for data files and exchange media for various types of fault, test, or simulation data for electrical power systems. It is intended to provide an easily interpretable form for use in exchanging data, but does not address encoding, compression, or transmission over networks.

IEEE C37.112
This standard defines certain characteristics of electro-mechanical and microprocessor-based overcurrent relays to coordinate constant current input and current conditions of different magnitudes.

IEEE C37.114
This guide outlines the techniques and application approaches to identify fault locations on AC transmission and distribution lines with an emphasis on automatic methods performed by computers. It reviews traditional approaches and the primary measurement techniques used in automated devices.

IEEE C37.115
This standard defines models, terminology, evaluation criteria, and performance measures for communication test scenarios between substation IEDs (Intelligent Electronic Devices) and with other local or remote applications. Communication protocols are not specified.

IEEE C37.116
Guide on protection and control issues for series capacitor bank installations in transmission facilities.

IEEE C37.117
Guide for use of protective relays in abnormal frequency load shedding and restoration.

IEEE C37.118-1-2011

A standard about synchrophasors and phasor measurement units for power systems. It defines synchronized phasors and frequency measurements and methods to verify measurements for analysis and operations. It replaces IEEE 1344.

IEEE C37.118-2-2011

A standard for realtime communications among phasor measurement units, phasor data concentrators, and other power systems applications.

IEEE C37.230

Guidelines for installation and coordination of relay systems on radial power system distribution lines. The guidelines describe distribution fundamentals, line configurations, and the advantages and disadvantages of different design approaches.

IEEE C37.231

Guidelines for the use and administration of firmware revisions for protection-related equipment to maximize the security and reliability of the power system.

IEEE C37.232

A substation standard that provides for naming TSD (Time Sequence Data) files, event sequences, and periodic logs. It was written to resolve problems associated with reporting, saving, exchanging, archiving, and retrieving large numbers of files. TSD refers to any electronic data file produced by digital relays, fault recorders, power swing recorders, or power quality monitors in which each data item corresponds to an instant of time that is identified by an explicit or implicit time tag. It is commonly known as COMNAME.

IEEE C37.238™
Standard that describes use of the precision time protocol identified in IEEE 1588 for automation, control, and protection of power system and data communication applications in Ethernet communications.

IEEE C37.239™
The Common Format for Event Data Exchange (COMFEDE) for Power Systems defines a common format for data files used to exchange event data collected from electrical power systems or power system models. It is a superset of what is in IEC 61850.

IEEE C57.13™
This is a series of six standards focused on instrument transformers, addressing electrical, dimensional, and mechanical characteristics and safety features for transformers used in generation, transmission, and distribution of alternating current.

IEEE C62.41
The recommended Practice in Surge Voltages in Low Voltage AC Circuits covers voltage and current tests to evaluate the surge withstand capabilities of equipment connected to utility grids. It addresses residential, commercial, and light industrial applications and has impacts on energy storage. It consists of the C62.41.1 guide and the C62.41.2 practice on the characterization of surges.

IEEE C62.45
This standard describes how to perform surge testing on electrical and electronic equipment using the recommended test waveforms defined in IEEE C62.41.2.

IEEE C63.022 - 1996
This standard identifies methods of measurement of spurious signals and radio disturbance emitted by information technology equipment. The

standard is applied in the context of protecting Smart Grid installations from unintentional EMF (electromagnetic frequency) interference, and is a republishing of IEC CISPR 22 as an American standard.

IEEE C63.16-1993
The American standard guide for electrostatic discharge test methodologies and criteria for tabletop and floor-standing electronic equipment.

IEEE PC37.118.1
Guidelines under development that specify frequency of synchrophasors and the rate of change of frequency measurements.

IEEE PC37.118.2
Guidelines under development that specify a method for power systems equipment to exchange synchrophasor measurement data. It identifies the types of realtime communications between PMUs and phasor data concentrators along with content and data formats.

IEEE PC37.236
Guidelines still in development that would define power system protective relay applications over digital communication channels, including operating and test procedures.

IEEE PC37.237
Recommended practices still in development that would define time tagging of power system protection events.

IEEE PC37.238
Standard still in development that defines the profile for use of IEEE 1588 Precision Time Protocol in Power System Applications. It describes the preferred physical layer and higher level definitions for precision time protocol methods and configuration guidelines.

IEEE PC37.239
Standard still in development that defines the common format for event data exchange for power systems, known as COMFEDE.

IEEE PC37.240
A proposed standard for cybersecurity requirements in substation applications, particularly for trust and assurance of data transmission and storage.

IEEE P1020™
The draft guidelines for the control of small hydroelectric generation that falls within 100 kVA to 5 MVA.

IEEE P1409
A proposed standard to define custom power and the guidelines and performance requirements for power quality and control.

IEEE P1547.5
Guidelines for interconnection of EPS greater than 10MVA to the power transmission grid

IEEE P1547.6
The guidelines for interconnecting distributed energy resources to urban distribution grids.

IEEE P1547.7
The proposed guide to conducting distribution impact studies for distributed energy resource interconnection.

IEEE P1547.8
A proposed standard that establishes methods and procedures for supplemental support of implementation strategies using IEEE 1547. It describes the communications required to support interconnection of

distributed energy resources to the grid, such as energy storage, intermittent renewables, and EVs.

IEEE P1595
A proposed set of standards to quantify the GHG credits attributable to Energy Efficiency and Renewable Energy projects.

IEEE P1642
This is a draft standard that establishes threat levels, protection methods, and monitoring and test techniques to protect computer systems and equipment from intentional EMI (Electromagnetic Interference).

IEEE P1711
A proposed standard that defines the cryptographic protocol to help implement cyber security to protect serial communications from cyber attacks.

IEEE P1777
This draft standard covers functional, performance, security, and on-site testing requirements for wireless technologies used in power system operations. Locations include substations, underground vaults, transmission and distribution circuits, generation and distributed energy resources plants, and customer electrical equipment

IEEE P1854
A standard in development that will categorize and describe important smart distribution applications including advanced automation and SCADA systems for reliability optimization, outage management, fault location and troubleshooting management, volt and var management, distributed resource and renewables integration, demand response, asset management, microgrids, and other applications.

IEEE P1909.1
A standard in development for test and installation procedures for communications equipment used in generation, transmission, and distribution. Tests cover safety, electromagnetic emissions, environmental, and mechanical factors.

IEEE P1901.2
A standard in development that covers PLC communications at frequencies below 500 KHz on AC and DC power lines. It covers communications for meters, EVs (electric vehicles) at charging stations, solar panels, and home area networks. It supports interoperability with PRIME and G3 PLC devices.

IEEE P1905.1
This standard in development for convergent digital home networks provides an abstraction layer to existing powerline, wireless, coaxial cable and Ethernet home networking technologies for fixed devices such as set top boxes, home gateways, and televisions, and mobile devices like laptops, tablets and smart phones.

IEEE P2030.1™
The renaming of an electric-sourced transportation standard previously known as P1809. It focuses on applications for electric vehicles and related support infrastructure, and delivers the foundation to address terminology, methods, equipment including charging systems, roadmap that includes privacy and roaming, and communications/cyber security for road transportation.

IEEE P2030.2™
A proposed standard for energy storage systems that defines interoperability characteristics and how different systems can be successfully integrated into the electric grid.

IEEE P2030.3™
A proposed standard for conformance test procedures for storage equipment and systems interconnection to the electrical grid.

IEEE P2030.4™
This standard is a guide for use of the smart grid interoperability reference model (SGIRM) within electric grid control and automation installations. It helps ensure the interoperability of Smart Grid devices and automation systems.

IEEE PES (IEEE Power and Energy Society)
A worldwide, nonprofit association with the mission to be the leading provider of scientific information on electric power and energy for the betterment of society and the preferred professional development source for its members. PES members are involved in the planning, research, development, construction, installation, and operation of equipment and systems for the safe, reliable, and economic generation, transmission, distribution, measurement, and control of electric energy. (http://www.ieee-pes.org/)

IEPA (Independent Energy Producers Association)
A California trade association representing the interests of developers and operators of independent energy facilities and independent power marketers. Members include producers of renewable products derived from biomass, geothermal, small hydro, solar, and wind; producers of highly efficient cogeneration; and owners/operators of gas-fired merchant facilities. (http://www.iepa.com/)

IERN (International Energy Regulation Network)
A web platform serving as an information exchange on electricity and natural gas market regulation for regulators and other energy market stakeholders. IERN is associated with the Council of European Energy Regulators (CEER). (http://www.iern.net/portal/page/portal/IERN_HOME)

IES (Illuminating Engineering Society of North America)
An industry association that shares information on lighting practices to its members and consumers. It promotes the exchange of ideas and information, and correlates research, investigations, and discussions to guide lighting professionals by consensus-based lighting recommendations. (http://www.iesna.org/)

IESO (Independent Electricity System Operator)
A not-for-profit ISO (Independent System Operator) established in 1998 by the Electricity Act of Ontario to manage the wholesale electricity market and oversee the reliable operation of Ontario's provincial electricity grid. Its members include almost 300 generators, traders, utilities, transmitters, and large-volume consumers, with 12.7 million consumers and 30 thousand kilometers of transmission lines. This ISO launched a Smart Grid Forum in 2009 to build consensus within industry and government about a vision for a future Smart Grid. It is one of the 10 ISO/RTOs currently operating in North America. (http://www.ieso.ca/)

IETF (Internet Engineering Task Force)
A large, open international community of network designers, operators, vendors, and researchers responsible for Internet standards. IETF designs and documents protocols such as IPv4 and IPv6, UDP, and 6LoWPAN, using RFCs (Requests for Comment) to build consensus and finalize standards. (http://www.ietf.org/index.html)

IFAS (Intelligent Feeder Automation System)
Multiple solutions that communicate with substation automation systems and distribution systems, including outage management, and can accept commands from the systems to deploy intelligence to the electrical grid.

IFMA (International Facility Management Association)
A global association for professional facility managers. IFMA certifies facility managers, conducts research, provides educational programs, including

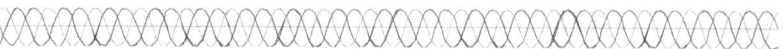

information about Smart Grid technologies to improve building energy consumption, and recognizes facility management certificate programs. (http://www.ifma.org/)

IGA (International Geothermal Association)
An international organization established to encourage R&D and utilization of geothermal resources worldwide through the publication of scientific and technical data and information, both within the community of geothermal specialists and between geothermal specialists and the public. It has more than 3,000 members in 65 countries. (http://www.geothermal-energy.org/)

IHD (In Home Display)
A dedicated device that presents electricity usage information, pricing information, and/or alerts to residential customers. It has a wired or wireless connection to a meter and may have a connection to a HEMS (Home Energy Management System) solution.

IIC (Industrial Internet Consortium™)
An industry association founded in 2014 by AT&T, Cisco, GE, IBM, and Intel to further the adoption of interconnected machines and analytics, and address the architectural framework for the Industrial Internet. Its goals include collecting best practices and case studies; facilitating input from academia, end users, vendors and governments; and influencing global standards requirements. (www.iiconsortium.org)

ILI (InLine Inspection)
A type of test for gas pipeline safety that uses PIGs (Pipeline Inspection Gauges) for periodic review of pipeline interiors and is part of a pipeline integrity management program.

ILI Association (InLine Inspection Association)
An industry association that promotes pipeline integrity through inline inspections by developing and proposing recommended practices and policies; promoting accreditation of the recommended practices under US and international standards; advocating policy at all levels of government; and member education. (http://www.iliassociation.org/)

Imbalance
A power disturbance generally caused by defective transformers or other malfunctioning equipment that creates a steady-state fault condition.

IMC (International M2M Council)
An industry association focused on publication of M2M business cases and ROI benchmarks across a number of industry verticals.   (http://im2mc.org)

IMM (Independent Market Monitor)
Entities that monitor and evaluate the performance of electricity markets in RTOs and ISOs and provide market oversight to help FERC (Federal Energy Regulatory Commission) ensure the competitiveness of wholesale electricity markets. Also known as MMUs (Market Monitoring Units).

Impedance
The opposition to power flow in an AC circuit or any device that introduces such opposition in the form of resistance, reactance, or both. Impedance is measured as the ratio of voltage to current, where a sinusoidal voltage and current of the same frequency are used for the measurement. Impedance is denoted in ohms.

Impoundment
A structure that contains water, such as a dam or embankment, to collect and store water for future use or control its flow.

IMT2000 (International Mobile Telecommunications 2000)
A set of ITU-approved, global standards commonly known as third-generation or 3G technologies. These technologies benefit from access to harmonized spectrum bands around the globe. The IMT-2000 provides a framework for worldwide wireless access by linking terrestrial- and satellite-based networks. The Radiocommunications Sector of the ITU received a proposal from IEEE to make a subset of the IEEE 802.16 or WiMAX standard part of IMT-2000, specifically the ITU-R technical standards. (http://www.itu.int/home/imt.html)

INCITS (InterNational Committee for Information Technology Standards)
An organization focused on development and maintenance of ICT (Information and Communications Technologies) standards. It covers storage, processing, transfer, display, management, organization, and retrieval of information and serves as ANSI's Technical Advisory Group for ISO/IEC Joint Technical Committee (JTC) 1. JTC 1 is responsible for International standardization in the field of Information Technology. (http://www.incits.org/)

Incremental energy costs
The additional cost of producing and/or transmitting electric energy above some previously determined base cost.

India Smart Grid Forum
A nonprofit public and private consortium of entities focused on accelerating the development of India's Smart Grid technologies. It is a member of the Global Smart Grid Federation.

Indirect utility cost
A utility cost that may not be meaningfully identified with any particular DSM (Demand-Side Management) program category. Indirect costs could be attributable to one of several accounting cost categories, such as administrative or marketing.

Induced seismicity
An earthquake caused by underground stresses resulting from injection of fluids into a drilling site.

Infiltration
The penetration of water from soil into sewer systems or other pipes through defective points, connections or manhole walls.

Influent
Water or wastewater moving into a storage facility (natural or manmade) or a treatment plant.

Information assurance
A term used in cyber security that describes the protection and defense of information and information systems, and includes restoration capabilities.

Information systems
Software and hardware that supports specific work operations, activities, and management decisions to improve organizational performance and quality of work life.

INGAA (Interstate Natural Gas Association of America)
A trade organization representing interstate natural gas transmission pipeline companies in the U.S. and Canada that advocates regulatory and legislative positions. (http://www.ingaa.org/)

Injection well
A well, pit, or hole that receives discharges of treated wastewater.

INL (Idaho National Laboratory)
An applied engineering national laboratory dedicated to supporting DOE (Department of Energy) missions in nuclear and energy research, science, and national defense. INL is operated for the DOE by Battelle Energy

Alliance. Research areas include the NSTB (National SCADA Test Bed) in collaboration with Sandia National Laboratory to analyze, test, and improve cyber security features in the control systems that operate the nation's electric power grid. INL also contains the Control Systems Security Center, funded by the Department of Homeland Security to identify and develop solutions to prevent cyber attacks of critical infrastructure, and it is also involved in energy storage research for transportation applications. (http://inl.gov)

InsideIQ Building Automation Alliance
A worldwide industry organization representing interests in building environment and energy services. Member firms participate in the development and deployment of building automation systems, energy services, and security platforms in their respective locations. (www.insideIQ.org)

Insolation
The amount of solar energy focused on a building or an area, expressed in annual KWh per square meter.

Integrated Grid
An EPRI (Electric Power Research Institute) initiative to incorporate DER (distributed energy resources) assets as an integral part of utility plans and operations of electricity grids and expand utility operational scopes to include DER operation. (http://integratedgrid.epri.com)

Intelligrid(SM)
An EPRI (Electric Power Research Institute) program designed to assist utilities through the process of Smart Grid technical foundation development, from conceptualization to deployment. The IntelliGrid initiative produced the IntelliGrid Architecture, an open standards, requirements-based approach for integrating data networks and

equipment for interoperability between different manufacturers' and utilities' systems. (http://intelligrid.epri.com/)

Interchange energy
Kilowatthours delivered to one electric utility or pooling system from another. Settlement may be payment, energy returned in kind at a later date, or accumulated as energy balances until the end of the stated period.

Interconnected network
A type of electrical distribution grid designed to provide high reliability to a group of customers. It consists of multiple wires interconnected with each other and a group of customers. This interconnection provides multiple paths on which electricity can flow from multiple power sources and offers the greatest reliability for continuity of service.

Interconnection
In the bulk power system, two or more electric systems with a common transmission or tie line to exchange electricity between them. Interconnections allow electricity to flow between the two grids and facilitate the sale of electricity between the two regions served by the systems. From a DG (Distributed Generation) perspective, it is the standards, utility practices, and policies that connect a DG resource to the grid.

Intermediate load
The range from base load to a point between base load and peak. This point may be the midpoint, a percentage of the peak load, or the load over a specified time period.

Intermittent electric generator
An electric generating plant that gets its energy from a variable source, such as solar, wind, or free-flowing rivers. Also known as an intermittent resource.

Internal demand
The total of metered outputs of all generators within the system, plus the metered line flows into the system, minus the metered line flows from the system. Auxiliary needs, such as equipment essential to operation of generating units, are not included in this total. Internal demand includes a calculation for all non-dispatchable demand response programs, such as real-time pricing, and some dispatchable demand response, such as demand bid/buyback.

International Downtown Association
A multinational organization whose goal is to connect diverse practitioners and professionals who work towards the improvement of cities into healthy, vibrant places. This is done through education, publications, advocacy, and advisory services. (www.ida-downtown.org)

Interoperability
From a communications perspective, the ability of multiple different systems, networks, devices, applications, or components to communicate, share and use information that has been securely exchanged without need for custom integrations or end user intervention. From a transmission or distribution grid perspective, the ability of multiple diverse systems, devices, and components to seamlessly and safely deliver electricity from generation to point of consumption.

Interruptible load
The customer load that can be interrupted at the time of annual peak load by the action of the consumer at the direct request of the system operator. This type of control usually involves large C&I (Commercial and Industrial) customers and is governed by a contract. There may be penalties for failure to curtail electricity consumption upon receipt of the request. Interruptible load does not include direct load control. It is also known as the curtailable load and is classified by NERC (North American Electric Reliability Corporation) as a form of dispatchable DR Demand Response).

Interruptible power
Power and associated energy made available by one utility to another that is subject to curtailment of delivery by the supplier in accordance with a prior agreement with the other party or under specified conditions.

Interruptible rate
The reduced price that customers receive in exchange for agreeing to reduce energy demand on short notice or allow the electric utility to temporarily cut off the energy supply. This interruption or reduction in demand typically occurs during periods of high demand for the energy. It is a form of demand response pricing that is most commonly applied to large C&I customers. Also known as the curtailable rate.

Interval metering
The measurement of customer energy use by fixed time periods or intervals. The interval time period is usually 15 minutes and is typically provided to C&I customers and some residential customers.

Inverter
A device that converts DC to AC electricity. This device is typically used to supply AC power from DC sources, such as solar panels, fuel cells, or batteries.

IoT (Internet of Things)
A conceptual description of the ability to connect any objects with an IP address and some level of embedded intelligence to the communications network. Embedded intelligence can include localization, sensing, identification, security, networking, processing, and control. IPv6 is a necessity for an Internet of Things.

IoT-A (Internet of Things – Architecture)
The European Lighthouse Integrated Project addressing the Internet-of-Things Architecture. Members include industry and universities focused on

developing evaluation frameworks for IoT projects, roadmaps, and an architectural reference model. (http://www.iot-a.eu/public/front-page)

IOU (Investor Owned Utility)
A type of utility that delivers power for retail consumption. It has publicly-traded shares and is regulated by a state utility commission.

IP500®
An open building automation platform. It utilizes wireless technology to communicate with a wide array of vendor-agnostic devices, as IP 500 supports the main industry norms and interfaces. Its primary function is to ensure interoperability between IP connected devices within buildings.

IP (Internet Protocol)
The protocol that identifies each connected device in a network with a unique address, and transmits information in packets that includes the destination address in each packet. It often appears as part of the TCP/IP acronym (*see* TCP/IP).

IP/MPLS (Internet Protocol/Multi Protocol Label Switching)
Existing communications technology used by utilities for its reliability and ability to prioritize communications to achieve latency targets for specific traffic.

IPC (Interphase Power Controller)
See IPFC.

IPFC (Interphase Power Flow Controller)
Technology that allows power to be transferred from one line to another in the high short-circuit environments commonly found in transmission grids. Devices with this technology replace thyristor controlled reactors and switched reactors plus SVCs (Static VAR Compensators).

IPMVP (International Performance Measurement and Verification Protocol)
An overview of current best-practice techniques available to energy projects in commercial and industrial facilities. It includes fuel-saving measures, water-efficiency measures, load shifting and energy reductions through installation or retrofit of equipment, and/or modification of operating procedures. It is used in ESCO (Energy Service Company) evaluations to assess the savings benefits of their energy performance contracts. (http://www.evo-world.org/index.php?option=com_docman&task=doc_view&gid=641)

IPP (Independent Power Producer)
An entity that owns or operates electric generation facilities to sell power to the public that is not an electric utility. Another term is non-utility generator (NUG).

IPRM (Interoperability Process Reference Manual)
This is a document developed by the SGTCC (Smart Grid Testing and Certification Committee) that details recommendations and best practices that enable the introduction of interoperable products in the marketplace, as well as enhances the adoption of consistent and measurable testing and certification methods across all Smart Grid standards.

IPSO Alliance (Internet Protocol for Smart Objects Alliance)
A global nonprofit organization that promotes IP as the basis for connecting smart objects or sensor and control networks (small computers with a sensor or an actuator and a communication device embedded in objects). Objects may include meters, building automation devices, and home appliances. It supports standards developed and ratified by the IETF (Internet Engineering Task Force), IEEE (Institute of Electrical and Electronics Engineers), or ISA (International Society of Automation). Members include meter and sensor manufacturers, AMI application developers, consultants, and utilities. (http://www.ipso-alliance.org/Pages/Front.php)

IPv6 (Internet Protocol version 6)
The latest version of IP, IPv6 uses 128-bit addresses, which increases the number of possible addresses to accommodate M2M devices and IoT applications. IPv6 allows for over 340 undecillion (340 trillion trillion trillion) IP addresses. IPv6 enables continued expansion of the internet into Smart Grid applications. The IPv6 protocols were developed by the IETF (Internet Engineering Task Force) and are seeing global deployment. It can be installed as a normal software upgrade in Internet devices and is interoperable with IPv4. It is also known as the IPng (Next Generation Internet Protocol). (http://datatracker.ietf.org/wg/ipv6/charter/)

IPv6 Forum
A global nonprofit industry association promoting IPv6 expertise, knowledge, applications, and interoperability. Its logos help identify tested applications and certified training. (http://www.ipv6forum.com/)

IRC (ISO/RTO Council)
An industry organization initiated in 2003, consisting of the ISO (Independent System Operators) and RTOs (Regional Transmission Organizations) in North America, serving two-thirds of electricity consumers in the United States and more than 50 percent of Canada's population. The IRC focuses on developing effective processes, tools, and standard methods to improve competitive electricity markets across North America. The IRC's goal is to balance electricity reliability considerations with market practices that result in electricity markets that provide competitive, reliable service to electricity users. One of its committees works with NERC (North American Electric Reliability Corporation) and NAESB (North American Energy Standards Board) to review wholesale electricity reliability and business standards (see M&V). Members are AESO, CAISO, ERCOT, IESO, ISO-NE, MISO, NBSO, NYISO, PJM, and SPP. (http://www.isorto.org)

IREC (Interstate Renewable Energy Council)
An organization whose mission is to accelerate the use of renewable energy sources and technologies in and through state and local government and community activities. It supports market-oriented services targeted at education, coordination, procurement, the adoption and implementation of uniform guidelines and standards, and consumer protection. Members include state and local government agencies, national laboratories, solar and renewable organizations and companies, and individuals. (http://www.irecusa.org/)

IRENA (International Renewable ENergy Association)
The mission of the nations that comprise membership is to promote the adoption and sustainable use of renewable energy. Renewables deployment is promoted through national policies and programs. (www.irena.org)

IRIG-B (Inter Range Instrumentation Group-B)
The most common version of a standard for time codes that allows equipment to be synchronized to a defined referent time. It is used in communication systems and telemetry systems that need time-of-day information to timestamp data. The other versions are -A, -D, -E, -G, and -H.

IROL (Interconnection Reliability Operating Limits)
A NERC (National Electricity Reliability Corporation) term that identifies voltage and time limits, that if violated, could lead to instability, or outages that negatively impact the reliability of the bulk electric system.

IRP (Integrated Resource Planning)
A process for evaluating rate recovery of planned utility projects that have benefits over time and in which externalities (ancillary benefits) are taken into consideration.

IRP (Interim Rate Policy)
A term used by FERC (Federal Energy Regulatory Commission) that defines criteria for cost recovery for public utilities to stimulate their Smart Grid investments. Criteria include that the investment must advance Smart Grid goals, it must consider cybersecurity and reliability requirements, it must adhere to open standards and upgradability, and information about the deployed investment must be shared with the DOE (Department of Energy). It is part of Docket No. PL09-4-000.

Irrigation
Controlled application of water to arable lands to supply water requirements not satisfied by rainfall. Powering irrigation consumes a great amount of electricity.

ISA (International Society of Automation)
A global, nonprofit organization involved in standards development for automation along with education, training, and certification services. The ISA 100 family of standards, focused on wireless systems for industrial automation, is working to define implementation procedures for wireless automation and control systems at the field level and will work with a variety of protocols. It was formerly known as the Instrumentation, Systems, and Automation Society. (http://www.isa.org/)

ISA99
A committee established to develop standards, recommend practices, and create reports to define and improve cyber security of industrial automation control systems, including SCADA systems.

ISA 100.11a
A wireless networking technology standard for industrial automation, specifically in the area of process control. ISA 100.11a uses IEEE 802 wireless technology to communicate with industrial plant factory automation and RFID applications, and is similar in use to WirelessHART.

ISACA (Information Systems Audit and Controls Association)
A global nonprofit association focused on the development, adoption and use of globally accepted, industry-leading knowledge and practices for information systems. It provides guidance, benchmarks and other tools that define the roles of information systems governance, security, audit and assurance professionals. (https://www.isaca.org)

ISF (Information Security Forum)
A not-for-profit association of international businesses that collaboratively share best practices, processes, and solutions in information security and risk management. (https://www.securityforum.org/)

(ISC)$^{2®}$ (International Information Systems Security Certification Consortium, Inc)
A global, not-for-profit organization that educates and certifies information security professionals. Certifications include CAP (Certified Authorization Professional); CISSP® (Certified Information Systems Security Professional); and CISSP concentrations for architecture (CISSP-ISSAP®), engineering (CISSP-ISSEP®), and management (CISSP-ISSMP®). (https://www.isc2.org/)

ISGAN (International Smart Grid Action Network)
An association of government entities, transmission and distribution system operators, national laboratories, research institutions, and power generators, to conduct a standard tracking of global Smart Grid activity. (http://www.cleanenergyministerial.org/our_work/smart_grid/index.html)

Islanding
A concept applied to distributed generation and integrated microgrids in which assets can disconnect and reconnect to the utility's distribution grid in a seamless and synchronized manner to isolate and contain disruptions to the larger grid. Islanding uses relays and controls to isolate loads, generating capacity, and energy storage to ensure that loads and resources stay in balance.

ISO (Independent System Operator)
An independent, federally-regulated (US or Canada) entity established to coordinate regional transmission in a non-discriminatory manner and ensure the safety and reliability of the electric system. These organizations monitor system loads and voltage profiles, operate transmission facilities and direct generation, define operating limits, develop contingency plans, and implement emergency procedures. ISOs also have authority over transmission expansion projects. This coordination, control, and monitoring of the electrical power system may be within a single U.S. state or across multiple states. There are currently 10 ISOs and RTOs (Regional Transmission Organizations) in North America.
(http://www.ferc.gov/market-oversight/mkt-electric/overview.asp)

ISO (International Organization for Standardization)
A non-governmental global developer and publisher of international standards, including ISO 9000 quality standards and ISO 14000 environmental management standards. (http://www.iso.org/)

ISO 65
A standard revised by ISO/IEC 17065:2012 that applies to third-parties that operate product certification systems. **ISO 772:1996**
A standard that gives terms, definitions and symbols in English and French languages that are used in the scope of hydrometric determinations or flow measurements, which are important for water conservation and management. It was amended in 2002 and 2004.

ISO 1452 1-5
A family of standards established in 2009 addressing plastic pipe for water supply, buried and above-ground drainage, and sewerage under pressure.

ISO 9506
An application layer specification for communications between devices and the applications that monitor and/or control them. Used as part of the MMS in industrial processes such as substation automation.

ISO 11298 1 & 3
Standards established in 2010 addressing plastic pipe for renovation of underground water supply networks.

ISO 11898-1
This standard specifies the DLL (data link layer) and physical signaling of the CAN (Controller Area Network). It provides support for setting up an interchange of digital information between modules implementing CAN DLL.

ISO 15693
This is a standard for vicinity contactless integrated circuit cards operating at a frequency of 13.56 MHz, which can be read from a distance of 1-1.5 meters, usually as identification cards. It includes parameters for initiating communications between card and reader, methods for communicating with multiple cards, and optimization of card detection.

ISO 15839
A water quality standard for on-line sensors and analyzing equipment for water. It includes specifications and performance tests.

ISO 16484-5
A building automation standard from BACnet that was adopted by the ISO (International Organization for Standardization). It addresses the design of new buildings and retrofit of existing buildings for acceptable indoor environment and practicable energy conservation and efficiency. Indoor environment includes air quality and thermal, acoustic, and visual factors. Formerly known as BACnet ANSI ASHRAE 135-2008.

ISO 22391 1-5
A family of standards established in 2009 that addresses plastic piping systems for hot and cold water installations using polyethylene of raised temperature resistance (PE-RT).

ISO 24510
Guidelines for the assessment and improvement of drinking water and wastewater service to users.

ISO 24511
Guidelines for the management of wastewater utilities and for the assessment of wastewater services

ISO 24512
Guidelines for the management of drinking water utilities and for the assessment of drinking water services.

ISO 27001
A standard for establishing, implementing, operating, monitoring, reviewing, maintaining, and improving an information management security system.

ISO 50001
A framework for energy performance improvements through best practices to reduce energy use of existing equipment and facilities; use energy performance data to take efficiency actions; and use of energy efficient equipment and systems.

ISO/IEC 12139-1
This standard is a physical and medium access control layer specification that conforms to the connectivity requirements of high-speed PLC (power line communication) stations operating below 30MHz.

ISO/IEC 14543-3-10 (International Organization for Standardization/International Electrotechnical Commission) A wireless short packet protocol for low power devices like energy-harvesting devices in a home application. It is optimized for energy harvesting solutions.

ISO/IEC 14908
This family of standards defines a communication protocol for local area control networks. It is suited to providing peer-to-peer communication for networked control and master-slave control strategies. They are the core standards used in LonMark verified products. Part 1 describes the services in the protocol stack, Part 2 specifies the control network protocol for twisted pair communication, Part 3 defines the power line channel specification, and Part 4 addresses requirements for IP communication.

ISO/IEC 15045
This standard outlines the residential gateway of an HES (Home Electronic System) that connects a home network to external network domains. It supports communications between devices within the home to service providers, operators, and other external users. It supports the connected home and applications for energy management, remote metering, smart appliances, and home security.

ISO/IEC 15067-3
This standard provides a model for a demand response energy management system for homes. It supports direct control, local control, and distributed control and supports DER (distributed energy resource) assets.

ISO/IEC 15118
This standard specifies the communication pathways between electric vehicles and their EVSE (Electric Vehicle Supply Equipment). Included in these pathways is vehicle detection within a Wi-Fi network for IP based

communication with charging equipment or other internet enabled devices such as smartphones.

ISO/IEC 17025:2012
A standard that specifies the general requirements for organizations certifying products, processes, and services. It is applicable to laboratories to develop management systems for quality, administrative, and technical operations.

ISO/IEC 17799
A standard that describes a code of practice for information security management. The latest version provides best practices, guidelines and general principles for implementing, maintaining and managing information security in any organization, producing and using information in any form. It covers critical success factors; organization of information security; asset management; human resources; physical and environmental security, and communications and operations management. It also includes consideration of information systems acquisition, development and maintenance; incident management; business continuity management; and compliance.

ISO/IEC 18000
A multi-part standard that provides a framework for common protocols for different frequencies for Radio Frequency Identification (RFID). The frequencies are: below 135 kHz, 13.56 MHz, 433 MHz, 860-960 MHz, 2.45 GHz, and 5.8 GHz. Common protocols encourage device interoperability and facilitate reduced implementation/operations costs for M2M communications in the Internet of Things.

ISO/IEC 18012
A two part standard that creates a framework for various components and defines a taxonomy and applications interoperability model for HES (Home Energy Systems) and building automation. It works with manual and

automatically configurable individual devices and groups of devices such as lighting, energy management, or home security controls.

ISO/IEC 21451-7
The standards family that covers a common set of data and control formats for sensors and actuators and is part of the standards composition for the Internet of Things. It is the same standard as IEEE 1451.

ISO/IEC 24752
A standard that describes user interface sockets with an XML-based language to facilitate operation of electronic products through remote or alternative interfaces and intelligent agents. It consists of five parts related to universal remote consoles that enable viewing of data and resulting actions by end users.

ISO/IEC 8824 ASN.1
This standard describes rules and structures for ASN.1 (Abstract Syntax Notation One). ASN.1 is used in encoding, transmitting, and decoding data sent over telecommunications networks. By creating a standardized notation, it becomes possible to automate the verification and validation of data presented in the ASN format.

ISO/IEC JTC1/SC 25 (ISO/IEC Joint Technical Committee 1/SubCommittee 25)
This is the joint ISO/IEC subcommittee responsible for creating standards for microprocessor systems, interfaces, and protocols for interconnecting media and information technology equipment. These standards are normally utilized in commercial and residential scenarios for embedded and distributed computing as well as input/output components.

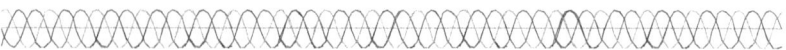

ISO/IEC JTC 1/SC 31 (ISO/IEC Joint Technical Committee 1/SubCommittee 31)
A collection of seven working M2M standards subgroups in Europe focused on automatic identification and data capture. Their goals are to define global interoperability protocols for wireless or air interfaces and data formats, as well as standards for RTLS (Real Time Locating Systems) and MIIM (Mobile Item Identification and Management) and security. Application standards can be based on this work.

ISO IWA 6:2008
Guidelines for a water utility, or any entity responsible for management of any part of a water supply system, to be prepared and ready to manage a water crisis.

ISO-NE (Independent System Operator – New England)
Established in 1997, this organization operates the region's power grid and wholesale electric markets, ensures the day-to-day reliable operation of New England's bulk power generation and transmission system, and manages comprehensive, regional planning processes. It covers the following states: Connecticut, Maine, Massachusetts, New Hampshire, Rhode Island, and Vermont. (http://www.iso-ne.com/)

ISO/TC 113
The ISO Technical Committee that develops standards for water measurement for use in water management and conservation. It has developed 75 standards.

ISO/TC 147
The ISO Technical Committee that develops standards for water quality, including definition of terms, water sampling, measurement and reporting of water characteristics. It has developed 246 standards.

ISO/TC 224
The ISO Technical Committee that developed three standards for the assessment, improvement and management of service activities relating to drinking water supply systems and wastewater systems.

ISSA (Information Systems Security Association)
A nonprofit, global association focused on the development, adoption and use of globally accepted, industry-leading knowledge and practices for information systems, specifically on information systems assurance, control and security, enterprise governance of IT, and IT-related risk and compliance. (https://www.isaca.org)

ITA (International Trade Administration)
A DOC agency that promotes trade and investment and ensures fair trade and compliance with trade laws and agreements. It manages the Market Development Cooperator Program that is a public/private partnership to promote solutions, including Smart Grid solutions, outside the USA. (http://trade.gov/)

ITCA (Interoperability Testing and Certification Authorities)
Entities responsible for testing interoperability standards proposed by organizations such as the SGIP (Smart Grid Interoperability panel). Once a proposed standard has been approved, ITCAs work towards implementation of the approved material.

ITSA (Intelligent Transportation Society of America)
Nonprofit organization that works with industry, state, and local planning agencies and the US Department of Transportation to develop intelligent transportation systems. Members include private corporations, public agencies, and academic institutions involved in the research, development, and design of smart surface transportation technologies that enhance safety, increase mobility, and sustain the environment. (http://www.itsa.org/)

ITU (International Telecommunications Union)
The leading UN-based agency for information and communication technology issues and the global focal point for governments and industry in developing networks and services. It coordinates the shared global use of the radio spectrum, promotes international cooperation in assigning satellite orbits, and establishes the worldwide standards for interconnection of communications systems. There are three sectors: radiocommunications, telecommunications standardization, and telecommunications development. The ITU is involved in research regarding sensor applications for monitoring power systems and remote sensing about weather conditions. The organization formed a Focus Group on Smart Grid in 2010 that reviewed and reported on Smart Grid standards in 2011. As a result of that work, the ITU formed a new group to coordinate Smart Grid and home networking standardization activities. (See JCA-SG&HN definition.) Membership includes 191 nations and more than 700 sector members and associates. (http://www.itu.int/net/home/index.aspx)

ITU Study Group 2
An ITU group responsible for the ITU T numbering standard E.164, which provides the structure and functionality for telephone numbers. It also works on ENUM, an Internet Engineering Task Force (IETF) protocol for entering E.164 numbers into the Internet domain name system (DNS). It has an important role in identify verification methods for M2M communications. (http://www.itu.int/net/ITU-T/info/sg02.aspx)

ITU-T
The standardization group of the ITU, focused on development of telecommunications standards. (http://www.itu.int/ITU-T/)

ITU-T Recommendations G.9955 (Annex B)
A standard that addresses physical layer specifications for narrow-band OFDM power line communications transmitters and receivers. It incorporates the PRIME PLC standard. It was approved as ITU-T G.9904.

ITU-T Recommendations G.9956 (Annex B)
A standard that addresses data layer specifications for narrow-band OFDM power line communications transmitters and receivers. It incorporates the PRIME PLC standard. It was approved as ITU-T G.9904.

IUN Coalition (Intelligent Utility Network Coalition)
A global industry group created by IBM in 2007 to help accelerate the global adoption of Smart Grid technologies and solutions. Activities include facilitating knowledge sharing, working with energy industry and standards groups, and using Intelligent Utility Network solutions and technologies.

IVVC (Integrated Volt/Var Control)
A system that reduces feeder network losses, maintains voltage profiles, and manages peak loads with feeder voltage reductions.

IWRM (Integrated Water Resources Management)
The planning practice that holistically views how water should be sustainably managed that addresses biosphere; hydrosphere; facilities; local, national and international policies; and uses.

IWWN (International Watch and Warning Network)
A 15-country consortium that identifies requirements for global network security and influences vendor behavior in the design and development of solutions. It was established in 2004 to foster international collaboration on addressing cyber threats, attacks, and vulnerabilities. Member countries are Australia, Canada, Finland, France, Germany, Hungary, Japan, Italy, the Netherlands, New Zealand, Norway, Sweden, Switzerland, the United Kingdom, and the United States.

J

J (Joule)
A measurement of work or energy, it is equal to the work done when a 1-ampere current is passed through a resistance of 1 ohm for 1 second.

JARI (Japan Automobile Research Institute)
An industry association that contributes to EV standards-setting bodies and conducts research on electric transport along with safety and environmental areas of interest. (http://www.jari.or.jp/english/)

JBEI (Joint BioEnergy Institute)
One of three DOE (Department of Energy) Bioenergy research centers that encourages collaboration among private industry, investors, and several national laboratories, including LBNL, SNLA, the University of California Berkeley and Davis campuses, the Carnegie Institution for Science, and Lawrence Livermore National Laboratory. (http://www.jbei.org)

JCA-SG&HN (Joint Coordination Activity on Smart Grid and Home Networking)
An ITU (International Telecommunications Union) group formed in 2012 to coordinate standards activity for Smart Grid and home networking with ITU study groups and other standards organizations.

JCESR (Joint Center for Energy Storage Research)
A US DOE R&D initiative for battery performance for mobile and fixed applications. It is also known as the Batteries and Energy Storage Hub, and is led by Argonne National Laboratory.

Jevon's Paradox
An economic hypothesis that states that as energy efficiency increases, so does demand or consumption of energy. In other words, the more efficiently you use a resource, the more of it you will use.

Joule's Law
The rate of heat production by a steady current in any part of an electrical circuit that is proportional to the resistance and to the square of the current. This law demonstrates the equivalence of mechanical and thermal energy.

JSCA (Japan Smart Community Alliance)
An organization that promotes public/private cooperative activities to realize smart communities and Smart Grid standardization. A member of the Global Smart Grid Federation.
(https://www.smartjapan.org/english/tabid/103/Default.aspx)

JTC (Joint Technical Committee)
Organizations with sub-committees (SC) that are formed by ISO and IEC to create international standards in areas that are in scope for both ISO and IEC interests.

Jurisdictional utilities
Utilities that are regulated by public laws.

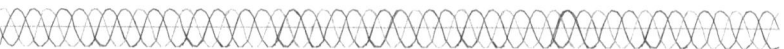

K

Kinetic theory of energy
The theory that the minute particles of all matter are in constant motion and that the temperature of a substance depends upon the velocity or speed of the motion.

Kirchhoff's Rules
Two laws that describe the flow of electricity. 1) The voltage rule: The sum of the voltage drops around a closed loop is equal to the sum of the voltage sources of that loop. Therefore, for any closed loop that can be traced in a circuit, the total voltage gained through EMF sources must equal the total voltage drops due to the presence of resistors, capacitors, or inductors in the closed loop. 2) The current rule: the charge that flows into a junction must be equal to the charge that flows out of that junction.

KNX®
A standard for home and building control, including lighting, security, heating, ventilation, air conditioning, water control, energy management, and metering. It supports wired and wireless communications media. It is approved as CENELEC EN 50090 and CEN EN 13321-1, ISO/IEC 14543-3, ANSI/ASHRAE 135, and GB/Z 20965.

KNX Association
An industry group that promotes the KONNEX standard for building management systems and offers testing and certification services. (http://www.knx.org/)

KSGA (Korea Smart Grid Association)
A member of the Global Smart Grid Federation focused on facilitating projects that establish a Smart Grid infrastructure, conduct Smart Grid R&D, mediate between government and public/private stakeholders, and establish a system for standardization. (www.k-smartgrid.org)

kVA (Kilovolt-Ampere)
A unit of apparent power, equal to 1,000 volt-amperes. It is the mathematical product of the volts and amperes in an electrical circuit. KVA is generally used to describe capacity or load.

kW (Kilowatt)
A unit of power, usually used for electric power. It is 1,000 watts. KW is generally used to describe demand.

kWh (Kilowatthour)
A unit of electrical energy, measured as 1 kilowatt of power expended for 1 hour. One kWh is equivalent to 3,412 Btu or 3.6 million Joules. Electricity rates are usually presented in cents per kWh.

L

L/A (Lead-Acid)
A common battery technology to store electricity. Lead-acid is one of the oldest and most developed battery technologies. It is a low-cost and common storage choice for power quality, UPS, and some spinning reserve applications. Its short cycle life limits its applications for energy management.

LAN (Local Area Network)
A multi-layer communications architecture that uses wired or wireless means to connect a group of devices. From an AMI (Advanced Metering Infrastructure) perspective, LANs extend from the gateway, router, or bridge between the WAN and the LAN to the end device (meter or control device).

Lat-Long (Latitude-Longitude)
A shorthand reference for latitude and longitude, commonly used in defining the location of assets.

Latency
A networking term for delays in transmitting data from one point to another, which can be LAN or WAN in scope. Latency increases response time in networks.

LBNL (Lawrence Berkeley National Laboratory)
A DOE (Department of Energy) national laboratory operated by the University of California that conducts energy research, among other topics. Three of its prominent research areas are energy efficiency, biofuels, and solar technology. Within energy efficiency, the lab has programs focused on developing more environmentally friendly technologies for generating and storing energy, including better batteries and fuel cells, better building energy efficiency, energy use and policy, and power-efficient computing. New funding initiatives include mathematical analyses related to the Smart Grid. (http://www.lbl.gov/)

LCFS (Low-Carbon Fuel Standard)
Established in the state of California by Executive order, this first-in-the-world GHG (Greenhouse Gas) standard for transportation fuels will help the state diversify its transportation fuel supplies, decrease the greenhouse gases emitted from those fuels, and establish a sustainable market for cleaner-burning fuels. (http://gov.ca.gov/index.php?/fact-sheet/5155/)

LCOE (Levelized Cost of Electricity)
A calculation that offers a common way to compare the cost of different energy technologies. It accounts for the installed system price and associated costs, such as financing, land, insurance, transmission, operation and maintenance, and depreciation, among other expenses. Carbon emission costs can also be taken into account, delivering a true comparison of all costs for different sources of energy.

Leachate
Contaminated water that leaks from landfills, dumps, fracking sites, or other manmade causes.

Leaching
A method to remove or extract material through use of water or another fluid percolation.

LEED® (Leadership in Energy Efficiency and Environmental Design)
A voluntary USGBC (US Green Building Council) program that promotes the design and construction of high-performance buildings to reduce or eliminate carbon energy consumption. It addresses all building types and emphasizes state-of-the-art strategies in five areas: sustainable site development, water savings, energy efficiency, materials and resources selection, and indoor environmental quality.
(http://www.usgbc.org/DisplayPage.aspx?CMSPageID=1988)

LER (Localized Energy Resource)
A term used in the California state Clean Energy Jobs Plan for onsite or small energy systems located close to consumption that avoid construction of transmission lines and use energy sources that typically do not have an environmental impact.

LESR (Limited Energy Storage Resource)
An energy storage asset that can be used for grid regulation because of its speed of response to dispatch. These assets have limited time durations, so are not used for long-term discharging.

Level 1 charge
The basic category of charging an EV (Electric Vehicle) or PHEV (Plugin Hybrid Electric Vehicle) uses the common 3-prong plug (single-phase AC with GFCI) for 120 V in North America. Level 1 charging requires between 8 to 30 hours to deliver a complete charge. Level 1 charging eliminates the need for upgrades to current electrical service, but the tradeoff is time to fully charge. It is typically used for smaller PHEVs and low-speed EVs.

Level 2 charge
This charge category uses the same 3-prong plug used in dryers and stoves to deliver 240-V single-phase AC power to EVs and PHEVs. Level 2 charging requires between 4 to 6 hours to deliver a complete charge. Level 2 charging delivers more convenient charging times, but the tradeoff is required upgrades to most electrical systems. It is generally recommended for larger PHEVs and EVs, and may be commonly deployed at homes, work places, and public charging stations.

Level 3 charge
This charge category delivers fast charging using 480-V 3-phase DC power. It requires dedicated EVSE (Electric Vehicle Supply Equipment) to deliver up to 50-percent battery replenishment within 10 minutes. Level 3 charging delivers the most convenient charging times, but requires dedicated power

supply equipment. It is most likely to be commonly deployed at public charging stations.

LFG (Landfill Gas)
The gaseous output, primarily CO_2 and methane, of organic waste in landfills. The methane from LFGs can be burned to produce electricity in a reciprocating engine or a turbine.

LGIA (Large Generator Interconnection Agreement)
A standard agreement used to secure the interconnection of 20-megawatt and higher generators to the grid. This agreement is part of the FERC (Federal Energy Regulatory Commission) Order 2003-C for ISOs (Independent System Operators), RTOs (Regional Transmission Organizations), and other transmission providers to use to govern interconnection.

LGIP (Large Generator Interconnection Procedure)
Procedures for the interconnection of generators larger than 20 megawatts. These procedures are mandated by FERC (Federal Energy Regulatory Commission) in Order No. 2003-C and designed for ISOs, RTOs, and other transmission providers to determine, in an equitable and transparent manner, the physical impacts to the transmission system and the associated investment requirements for interconnecting new large generators to the grid.

Li-ion (Lithium ion)
An energy storage technology that has high-energy density, almost 100-percent efficiency, and a long cycle life. Its main limitation is the cost involved to engineer safety into the batteries.

LIDAR (Light Detection And Ranging)
An aerial planning and/or inspection method for overhead transmission cables.

Line loss
Energy loss resulting from transmission of electrical energy across power lines. The losses generally occur within transmission systems, but can also occur in distribution systems. These losses occur due to the conversion of electricity to heat and electromagnetic energy.

Line voltage
The voltage measured between two conductors or phases in a three-phase system.

Linear Fresnel reflector system
A type of concentrated solar power for utility-scale generation that uses an array of flat or slightly curved reflectors to collect and concentrate solar energy to heat oil flowing through the pipe. The heat energy is then used to generate electricity in a conventional steam generator.

Lm/W (Lumens/Watt)
A measurement often used with wattage to determine the luminous efficiency of a light-emitting body.

LMP (Locational Marginal Pricing)
A FERC (Federal Energy Regulatory Commission)-recommended pricing approach used to manage the efficient use of the electric transmission system during periods of congestion. It provides market participants accurate realtime price signal of electricity at every location on the grid. These prices, in turn, reveal the value of locating new generation, upgrading transmission, or reducing electricity consumption through demand response programs. It is currently in use in PJM's wholesale electricity markets.

Load
The definition of load is distinguished by its context. From a bulk power perspective, it is the moment-to-moment amount of electric power

required based on the consumption pattern of customers. It can also refer to demand over a period of time instead of a single moment in time, such as industrial load, commercial load, or residential load. It can also refer to any electrical device consuming energy in a circuit. From a demand-side perspective, load is the capacity that is committed to previously agreed load reductions under certain conditions.

Load control program
A program in which a utility offers a lower electricity rate in exchange for authorization to turn off participating customers' air conditioners or water heaters for short periods of time by remote control to reduce peak demand. Also known as demand response.

Load curve
The distribution of power over time, usually illustrated in a graph with time as the horizontal axis and load as the vertical axis.

Load diversity
The difference between the peak of coincident and non-coincident demands of two or more individual loads.

Load factor
The ratio of average load to peak load over a period of time. It is measured in kilowatts and describes the effective utilization of distribution system components.

Load following
Monitoring the power output of electric generators within a utility's distribution area to respond to changes in system frequency, tieline loading, or other variables. Utilities may add additional generation to maintain the scheduled system frequency and/or exchanges with other areas. This practice ensures that generators are producing the right amount of energy to supply to utility customers.

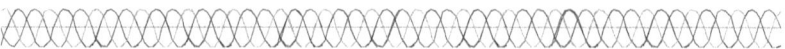

Load leveling
Any load-control technique that reduces the cyclical daily load flows and increases baseload generation. Peak load pricing and time-of-day charges are two techniques that utilities use to reduce peak load and to maximize efficient generation of electricity. It can also refer to the production of energy during off-peak periods for storage and use during peak demand periods.

Load management
Practices that reduce the maximum demand by different classes of customers to help the utility meet system capability for a given period of time. Some practices, such as using stored off-peak power during peak periods, may become permanent rather than used a few events a year. Other practices, like peak shifting, get customers to schedule their peak demands at times that do not overlap and stress the system.

Load pocket
An area where there is insufficient transmission to reliably supply 100 percent of the electric load, requiring generation capacity that is physically located within that area. It is typically found in areas with high concentrations of intensive power use, like urban areas.

Load profile
A visual display of the amount of power consumed, usually on an hourly basis. A load profile at the bulk power system is an accumulation of all demand. At an individual consumer level, a load profile might be displayed in shorter time increments, like 15 or 30 minutes, and provide additional analysis of consumption patterns.

Load reduction request
Requests to any customer or the general public to reduce the use of electricity to ensure the continuity of service of the utility's bulk electric power supply system.

Load shape
A graphical method of describing load and the relationship of power supplied to the time of occurrence. From a DSM perspective, it refers to the distribution of energy requirements over time.

Load shedding
Intentional actions by a utility to reduce the load on the system. These are usually conducted during emergency periods, such as capacity shortages, system instability, or voltage control, and have different impacts on different classes of customers.

Load shifting
Demand-side management programs that motivate customers to shift electricity use from peak to off-peak times.

Load zone
A geographical territory in a utility that has a specific price for electricity.

Loading order
The California electricity resources policy that calls for meeting new electricity needs by increasing energy efficiency and demand response, and meeting new generation needs first with renewable and distributed generation resources and second with clean fossil-fueled generation. The loading order concept was originally adopted in the 2003 California Energy Action Plan I, a collaborative effort by the CEC (California Energy Commission) and the CPUC (California Public Utilities Commission). (http://www.energy.ca.gov/2005publications/CEC-400-2005-043/CEC-400-2005-043.PDF)

Lockbox
A cyber security term describing how encrypted data is sent from device to device. The main premise is that both devices require a special "key" and certificate of authority to decrypt the encrypted information.

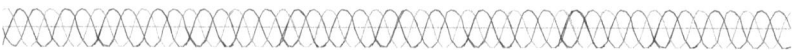

Logic bomb
Software code that is dormant until activated by a specific event or action. Smart Grid cybersecurity standards must protect against a threat like this.

Logical interface
As defined in the ARRA Section 1306 amending EISA 2007 (Energy Independence and Security Act of 2007), the standard functional interface specification that enables electric equipment or appliances to receive and act on signals from any intelligent system in the Smart Grid, such as a demand response system.

LOLE (Loss of Load Expectation)
The measure of the number of hours per year, days per year, or days in 10 years that generation may be inadequate to serve load.

LOLP (Loss of Load Probability)
A measure of the probability that system demand will exceed capacity during a given period. It is usually stated as the estimated number of days over a long period, frequently 10 years or the life of the system.

LOLP-weighted peak (Loss of Load Probability-weighted peak)
Demand that is normalized by relative LOLP calculations from production cost models. It is one type of measure of peak capacity that is used to calculate the allocation of costs to customer classes in regulatory cost of service models.

LonMark® International
An international industry consortium that promotes development, specification, and use of products using ISO/IEC 14908-1 and related standards and LonWorks control networks. Products that are verified to conform to the LonMark interoperability guidelines are eligible to carry the LonMark logo. These products are found in building automation today and

may be found in home automation applications in the future. (http://www.lonmark.org/)

LonWorks

A network platform that consists of a group of devices networked together for sensing, monitoring, communication, and control, connected by routers that communicate to one another using a common protocol. It is generally known as a control network, and its applications in building automation are based on standards approved by ISO/IEC, ANSI (American National Standards Institute), and others. Control networks can include industrial process management, building automation systems, and home and utility controls.

Loop flow

The movement of electric power from generator to load by dividing along multiple parallel paths, usually referring to power flow along an unintended path that loops away from the most direct geographic path or contract path.

Looped topology

A network configuration in which a number of devices or nodes are serially connected, and the last node is connected to the first node. Also known as a ring topology.

LoS (Line of Sight)

A radio frequency term that refers to the placement of transmitting and receiving antennas so that there are no obstructions to the radio waves between the antennas.

Loss of service

Any loss of electricity to customers for a defined period of duration or affecting a defined number of customers. These definitions may vary for different regulatory agencies. Also known as load loss.

Low-Voltage DC (LVDC) Forum

An IEEE Standards Association forum based in India that is focused on demonstration projects that contribute knowledge to low voltage standards and commercialization of supporting technologies. Projects include products, systems and solutions up to 1100V DC.

LVDC (Low Voltage direct current)

DC electricity operating in the range of 120-1500 Volts. LVDC is used in small scale microgrids such as singular homes and is directly operable with DC produced by PV and other renewable energy generation systems.

LPPC (Large Public Power Council)

An organization that comprises 23 of the nation's largest locally owned and controlled, not-for-profit power systems. LPPC members represent 90 percent of the public-agency owned, but non-federal, transmission investment in the nation and work to develop and advance consumer-oriented positions on national energy issues. (www.lppc.org)

LSE (Load-Serving Entity)

Any entity, including a load aggregator or power marketer, that has the authority or obligation through law, regulation, franchise, or contract to serve and sell electric energy to end users located within a defined territory.

LTE (Long-Term Evolution)

A 4G, IP-based wireless platform that converges GSM (Global System for Mobile Communication™) and CDMA (Code Division Multiple Access) technologies. It is backwards compatible with 3G. It outpaces CDMA performance, particularly in larger channel bandwidths. Its peak download rate is at least 100 Mbps and the maximum upload rate is at least 50 Mbps.

Lumen
The standard unit for the luminous flux of a light source, it is "quantity" of light emitted by the light source. It identifies the energy per unit of time that is radiated from a source over wavelengths sensitive to the human eye.

Lumen Coalition
An adhoc consortium of manufacturers, retailers, utilities, energy efficiency groups and other organizations focused on consumer education about lighting solutions. (http://lumennow.org/)

LVRT (Low Voltage Ride-Through)
The ability of wind turbines to continue operations during network disturbances such as voltage drops using capacitor banks and battery backup. It is usually required for grid integration. Also known as Fault Ride-Through or FRT.

M

M & V (Measurement and Verification)
Generally accepted guidelines to measure and verify the load impacts of demand response resources. There is a lack of standards now, but the NAESB (North American Energy Standards Board) is developing standards for wholesale and retail markets to reduce transaction costs and barriers to participation.

M-Bus (Meter Bus)
A European standard for two-way meters.

M-Class type synchrophasor
Devices dedicated to capturing measurement data instead of protection data. Measurement data does not have the same time criticality that protection data has, and can be sent by slower signaling paths.

M/441
A European Commission standardization mandate to CEN, CENELEC, and ETSI for development of an open architecture for utility meters' communications protocols to enable interoperability. It applies to water, gas, electricity, and heat meters.

M/490
A European Commission mandate for a fully integrated and interoperable Smart Grid. It is also known as the Smart Grid Mandate and the Standardization Mandate to European Standardization Organizations to support European Smart Grid deployment. Building, industry, appliances and home automation are out of scope of this mandate.

M2M (Machine to Machine)
Wired or wireless IP-based communications between fixed and mobile devices that can transmit and/or receive data, and can optionally take actions without human intervention. From a Smart Grid perspective,

communication between devices in a grid that could range from simple condition-based sensing to complex condition-based participation in markets.

MADRI (Mid-Atlantic Distributed Resources Initiative)
An organization that seeks to identify and remedy retail barriers to the deployment of distributed generation, demand response, and energy efficiency in the mid-Atlantic region. It was established in 2004 by the public utility commissions of Delaware, District of Columbia, Maryland, New Jersey, and Pennsylvania, the DOE (Department of Energy), EPA (Environmental Protection Agency), FERC (Federal Energy Regulatory Commission), and PJM Interconnection.
(http://sites.energetics.com/madri/)

MAGICC (Mid-Atlantic Grid Interactive Car Consortium)
An organization of industry and academic entities developing, testing, and demonstrating V2G technologies. (http://www.magicconsortium.org/)

Magnetometers
A type of sensor that detects direction.

MAIFI (Momentary Average Interruption Frequency Index)
A calculation of the average frequency of momentary interruptions. Its formula is the total number of customer momentary interruptions divided by the total number of customers served.

MAN (Metropolitan Area Network)
An interconnected network of LANs with a high capacity backbone that is usually built to serve cities or campuses.

Man-in-the-middle attack
An exploit where an attacker actively eavesdrops on a communication channel between two parties by independently connecting with each party

and then relaying messages between them, leading the parties to believe they are directly communicating with each other over a private and secure channel. Because the attacker has full control of the conversation, he can inject new messages into the discussion, unbeknownst to the parties.

MAPL (Maximum Allowable Path Loss)
The amount of signal power loss or attenuation that is acceptable in a wireless network.

MAPP (Mid-continent Area Power Pool)
The former name for the organization now known as the Midwest Reliability Organization, or MRO.

Master meter
A meter that collects electricity use for multi-tenant dwellings. The master-meter point of contact receives a single bill from the utility and collects from tenants individually.

MCFC (Molten-Carbonate Fuel Cell)
Fuel cells applicable for large stationary power generation and co-generation facilities. Characteristics include operation around 600 degrees Celsius and the use of more economical materials than SOFCs (Solid Oxide Fuel Cell). The high temperatures produce steam that can be channeled into turbines to generate more electricity, also known as co-generation.

MCI (Modular Communications Interface)
A specification based on work from USNAP and EPRI) that will allow consumer products manufacturers to furnish Smart Grid ready products capable of getting energy information from digital meters and energy system interfaces with any communications technology.

MCL (Maximum contaminant level)
Standards outlining the allowable levels of contaminants in public water systems established by the EPA (Environmental Protection Agency).

MDA (Mobile Data Association)
An industry association focused on all mobile data centric businesses. Members include mobile network operators, content aggregators, phone manufacturers, and application and content developers. (http://www.themda.org/index.php)

MDM (Meter Data Management)
A system that manages data from meter readings, usually for billing and historical analysis.

MDMS (Meter Data Management System)
A software system that can gather data from different AMI or metering systems and provide that data to other applications, such as revenue protection, load research, billing, forecasting, customer service, system operation, and maintenance.

MFL (Magnetic Flux Leakage)
A type of ILI (InLine Integrity) test that finds pipeline defects by applying a saturating magnetic field, supplied by huge magnets, into the pipe material and then sensing any local changes in this field. Testing sensitivity ranges from standard, low-resolution to high and extra high resolution, and is based on the quantity of sensors in the PIG (Pipeline Insertable Gauge).

Measurement Canada
The governmental organization with sole jurisdiction for ensuring the integrity and accuracy of measurement in the Canadian marketplace. The Electricity and Gas Inspection Act and Regulations set the rules for purchase and sale of electricity and natural gas and include approval for the electric

and gas meters used in Canada. (http://www.ic.gc.ca/eic/site/mc-mc.nsf/eng/home)

Mechanical Systems and Control Group
Part of NIST (National Institute of Standards and Technology), this group promotes development and deployment of building mechanical systems and controls; develops tools, diagnostic procedures, and performance evaluation techniques to quantify the performance of building HVAC equipment and systems; develops standard communication protocols between building management and control systems; and develops the technical bases for advanced building controls. It contains two working groups that have impacts on Smart Grid activities: the BACnet Utility Integration Working Group and the BACnet Network Security Working Group.

Mega grid
A future grid that will provide links between today's regional grid layouts and planned renewable energy generators to transmit electricity to any region where needed. The vision for this grid is also to eliminate congestion problems and balance loads from intermittent energy sources across regions. Also known as a super grid or national grid.

MEL (Miscellaneous Energy Loads)
Electricity uses within a building that are not the "traditional" three electricity consumers; lighting, refrigeration, or HVAC. Essentially any device that is plugged in to a socket is considered a MEL. A major contributor to MEL is electronic devices with "standby" modes that are constantly consuming electricity.

MEMS (MicroElectroMechanical Systems)
A fabrication approach for miniaturized mechanical and electro-mechanical devices and structures that perform a mechanical functionality. The size of a MEMS device can range from below one micron to several millimeters,

and should not be confused with nanotechnology. Transducers and actuators are examples of MEMS.

Mesh network
This wireless communications network topology spreads nodes that communicate with each other across a territory. The nodes are small radio transmitters that use the common WiFi standards known as 802.11a, b and g to communicate with devices and each other. Communications software defines the protocols for how nodes interact, and dynamic routing lets nodes automatically choose the best path to a receiver. This topology can be useful where there are LoS (Line of Sight) issues.

Metadata
Information that describes the attributes of data, such as size of a document, authorship, image resolution, or other pertinent content details.

Meter
A utility device that measures and records electricity use, retrieved by manual means, to calculate customer bills. The devices may be solid state or electromechanical. It is sometimes referred to as the utility's cash register.

MGA (Microgrid Alliance)
An industry organization working to accelerate the growth and development of microgrids. This is done through collective contributions to microgrid-friendly policy development and microgrid-related businesses. (http://www.microgridalliance.org)

MGS (Modern Grid Strategy)
An initiative focused on developing a vision for a modern grid, sharing information, and coordinating regional technology integration projects through industry/DOE (Department of Energy) partnerships that invest in demonstrations of the modern grid vision. It is managed by NETL (National

Energy Technology Laboratory).
(http://www.netl.doe.gov/moderngrid/index.html)

Microgrid
A small power system that integrates self-contained generation, distribution, sensors, energy storage, and energy management software with a seamless and synchronized connection to a utility power system, and can operate independently as an island from that system. Generation includes renewable energy sources and the ability to sell back excess capacity to a utility. On-site microgrid management software includes controls for the power generation, utility connect/disconnect, distribution, and energy storage equipment along with building energy management applications for industrial, commercial, or home use. CERTS (Consortium for Electric Reliability Technology Solutions) has documented a microgrid concept.

Microturbine
From a distributed-energy perspective, a small combustion turbine that produces between 25 kW and 500 kW of power. There are two general classes of microturbines: those that recover the exhaust heat to improve overall efficiency (recuperated) and simple cycle generators that do not recover heat.

Middleware
Software that enables other software applications to exchange data and achieve interoperability. It can include application programming interfaces or custom code.

Midwestern Greenhouse Gas Reduction Accord
A regional agreement signed by Minnesota, Wisconsin, Michigan, Illinois, Kansas, Iowa, and the Canadian province of Manitoba to reduce GHG emissions. The states of South Dakota, Indiana, and Ohio have observer status.

MIIM (Mobile Item Identification and Management)
A term used in standards groups regarding automatic identification and data collection techniques for mobile devices that connect to wired or wireless networks, including sensor specifications; combining RFID (Radio Frequency Identification) with mobile devices; and combining optically readable media with mobile devices. It is Working Group 6 in the ISO/IEC JTC 1/SC 31.

MirrorLink^{TM}
A technology standard promoted by the Car Connectivity Consortium that offers common controls for accessing car radio, climate controls, and navigation systems in cars.

MISO (Midwest Independent System Operator)
Established in 2002, this ISO/RTO entity optimizes the efficiency of the interconnected system, provides regional solutions to regional planning needs, and continually minimizes any risk to reliability. It administers a two-settlement (day-ahead and real-time) energy market known as the Day-2 market. It produces hourly locational marginal prices that are rolled up into five regional hub prices. MISO also administers a monthly FTR (Financial Transmission Rights) allocation and auction. Its territory includes all or most of North Dakota, South Dakota, Nebraska, Minnesota, Iowa, Wisconsin, Illinois, Indiana, Michigan, and parts of Montana, Missouri, Kentucky, and Ohio. It is one of the 10 ISO/RTOs currently operating in North America. (http://www.midwestiso.org/home)

MJU (Multi-Jurisdictional Utility)
Utility that covers more than one state. It could be an IOU (Investor Owned Utility), municipality, or co-op structure.

MMDS (Multi-channel Multi-point Distribution System)
A commercial broadband wireless point-to-multipoint specification using UHF (Ultra High Frequency) communications. MMDS operates on FCC

(Federal Communications Commission)-licensed frequencies. It is now known as Broadband Radio Service (BRS).

MMS (Manufacturing Message Specification)
An application layer protocol that specifies services for the exchange of SCADA data between devices and applications in industrial processes. It is part of ISO 9506 and is mapped to IEC 61850. Also plays a role in IoT communications for intelligent electronic devices.

MMU (Market Monitoring Unit)
Entities that monitor and evaluate the performance of electricity markets in RTOs and ISOs and provide market oversight to help FERC (Federal Energy Regulatory Commission) ensure the competitiveness of wholesale electricity markets. Also known as an IMM (Independent Market Monitor).

MNO (Mobile Network Operator)
A provider of wireless/cellular communications services with access to all hardware and systems necessary for providing voice and data services to end users. MNOs own wireless network infrastructure, radio spectrum allocation, and customer management and engagement systems.

MNVO (Mobile Network Virtual Operator)
A communications company that owns the mobile customer relationship, but does not necessarily own the network providing the wireless services. The network capacity may be leased from an MNO (Mobile Network Operator).

MoCA® (Multimedia Over Coax Alliance)
An industry consortium that promotes MoCA technology as the worldwide standard for home entertainment networking. It is used by all three pay TV segments—cable, satellite and IPTV in any of three specification forms. It is a DLNA-approved layer two protocol for their Interoperability

Guidelines and is an abstraction layer for transport protocols such as WiFi, Ethernet and HomePlug in IEEE 1905. (http://www.mocalliance.org/)

Mobile radio
A wireless narrowband communications medium commonly used by utilities' private networks to provide coverage between remote assets in transmission and distribution networks and/or mobile workforce applications.

MODBUS
A de-facto standard communications protocol published by Modicon (now Schneider Electric) in 1979. It is used to establish master-slave/ client-server communication between intelligent devices such as sensors and field devices in industrial and commercial applications.

Monitoring well
A well designed to measure water levels, water quality, and/or seepage of detectable substances.

Monthly peak
Average of the single highest demand occurring each month for 12 months. It is one type of measure of peak capacity that is used to calculate transmission reservation costs.

Moore's law
A prediction made by Gordon Moore, Intel co-founder, about doubling transistor density on integrated circuits about every two years that is now commonly accepted as fact. Microprocessors progressively became smaller and more powerful as a result.

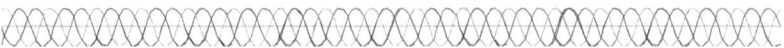

MPLS (Multiprotocol Label Switching)
A mechanism that creates a short fixed-length label that is shorthand for the entire IP packet header and is useful in large-scale IP, ATM, Ethernet, and Frame Relay networks.

MPR (Market Price Referent)
A CPUC (California Public Utilities Commission) calculation based on the long-term ownership, operating, and fixed-price fuel costs for a new 500-MW natural gas-fired combined-cycle gas turbine. It is used to judge the cost-effectiveness of renewable energy projects under the current renewable energy mandates.

MQTT
This is a lightweight messaging transport protocol developed by OASIS and M2Mi for the M2M and IoT marketplace. MQTT is a publish/subscribe messaging transport protocol to connect sensors, actuators, and smart mobile devices such as phones and tablets using already installed software processing technology.

MRC (Microgrid Resources Coalition)
A consortium of microgrid owners, operators, developers, and suppliers advocating for microgrids as accepted energy resources. This is accomplished through legal advocacy, encouraging the creation of microgrid-friendly regulations and tariffs, and market support. (http://www.microgridresources.com/)

MRL (Minimum Reporting Level)
The smallest measured concentration of a substance that can be reliably measured by using a given analytical method, used in water quality assessments.

MRO (Midwest Reliability Organization)
A nonprofit organization formed in 2002 that is dedicated to ensuring the reliability of the bulk power system in the north central region of North America. MRO is one of the eight NERC (North American Electric Reliability Corporation) Regional Entities or reliability regions. It assumed the reliability functions of the Mid-Continent Area Power Pool (MAPP) and Mid-America Interconnected Network (MAIN). Members are investor-owned utilities, cooperatives, municipals, public power districts, a power marketing agency, power marketers, regulatory agencies, and independent power producers from the following states and provinces: Minnesota, Nebraska, North Dakota, Manitoba, Saskatchewan, and parts of Wisconsin, Montana, Iowa, and South Dakota. It serves over 16 million people and covers nearly 1 million square miles.
(http://www.midwestreliability.org/about_mro.html)

MRTU (Market Reform and Technology Update)
A CAISO electric market structure launched in March 2009 with the objectives to enhance grid reliability and fix flaws in the ISO (Independent System Operator) market. It keeps California compatible with market designs that are working throughout North America and replaces aging technology with modern computer systems. The redesign introduces a day-ahead market for energy that helps firm up next-day production and delivery schedules and enable grid operators to manage transmission bottlenecks efficiently so that electricity flows without interruption.

MSS (Manufacturers Standardization Society of the Valve and Fittings Industry)
A nonprofit technical association that promotes industry, national and international codes and standards for valves, pipe fittings, flanges, and associated seals. It works with American Society of Mechanical Engineers (ASME), American National Standards Institute (ANSI), American Society for Testing and Materials (ASTM), American Waterworks Association (AWWA)

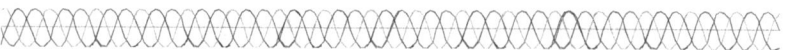

and the National Fire Protection Association (NFPA). (http://mss-hq.org/Store/AboutUs.cfm)

Multi-utilities
Utilities that provide water and other services such as gas and/or electricity.

Multicast
A one-to-many transmission method in which one transmitter sends a message to multiple recipients or receivers.

Multicore technology
Computer chips with two or more processor chips or cores working simultaneously as one system. It is a technology that improves computing-energy efficiency.

MultiSpeak®
A joint initiative from NRECA (National Rural Electric Cooperative Association) utilities, and software vendors that created a specification for data exchange interfaces, with certification provided by a third party testing facility. It defines what data needs to be exchanged between common software, defines the structure of those data objects, specifies message structures and messaging architectures, and supports web services and/or sockets and SOAP data exchanges. It serves investor-owned, municipal, and cooperative utilities, but is scalable to any size electricity provider. Interfaces include meter reading, connect/ disconnect, meter data management, GIS (Geographic Information System), SCADA (Supervisory Control and Data Acquisition), DR (Demand Response) and distribution automation control. (http://www.multispeak.org)

MultiSpeak V4.X
The most recent updates contain support for CIM (Common Information Model) and CPSM (Common Power System Model) applications.

Municipal Utility
A not-for-profit utility that is owned by a community or local government that furnishes electricity, gas, and/or water to that community. Also known as munis.

MUSH market (Municipalities, Universities, K-12 Schools & Hospitals market)
A collection of entities that have common characteristics from an energy services perspective.

MVNO (Mobile Virtual Network Operator)
A company that buys network capacity, such as minutes of use, from a network operator that owns bandwidth to brand and sell mobile services to customers.

MW (Megawatt)
A standard measure of electric power plant generating capacity that is equivalent to 1000 kilowatts or 1 million watts.

MWDRI (Midwest Demand Resources Initiative)
A working group within MISO that promotes progress toward active demand-side programs throughout its region by market design, pricing, and technology. (http://www.misostates.org/WG8MWDRIList.htm)

MWh (Megawatt Hour)
The basic industrial unit for pricing electricity, equal to 1,000 kilowatts of power supplied continuously for 1 hour.

MW$_{th}$
Megawatts of thermal energy produced.

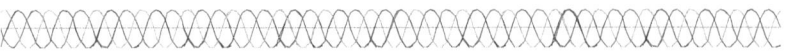

MYPP (Multi-Year Program Plan)
Operational guides for programs to manage their activities and for government departments, such as the DOE (Department of Energy), to track progress toward goals. An MYPP spans five years and includes future plans that may or may not be funded.

N

NACAA (National Association of Clean Air Agencies)
An organization that represents air pollution control agencies in 53 states and territories and over 165 major metropolitan areas across the USA (http://www.4cleanair.org/)

NACo (National Association of Counties)
An organization that represents county governments in policy matters and provides education and research programs. Members include more than 2,300 counties (of the 3,068 counties in the USA), representing more than 80 percent of the population. It offers a range of services to help county officials protect water resources on the local level. (http://www.naco.org/about/who/Pages/default.aspx)

NACWA (National Association of Clean Water Agencies)
An organization representing the interests and priorities of publicly-owned water treatment works. It promotes the development and implementation of scientifically based, technically sound and cost-effective national environmental programs and policies for clean water. (http://www.nacwa.org/)

NAESB (North American Energy Standards Board)
An industry forum for the development and promotion of standards to lead to a seamless marketplace for wholesale and retail electricity and natural gas. NAESB works with NERC (North American Electric Reliability Corporation) to ensure tight integration of their respective standards development processes where reliability and commercial needs are closely related. It is currently developing business practice standards for the measurement and verification of demand reductions from wholesale and retail demand response programs. NAESB used to be known as GISB (Gas Industry Standards Board). (http://www.naesb.org/default.htm)

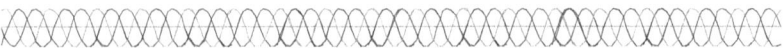

NAESB REQ21 (North American Energy Standards Board Requirement 21) Energy Services Provider Interface (ESPI)
This is an interface standard for energy service providers. Its primary function is to create a uniform process for the exchange of a retail customer's energy usage information and the distribution company. This is open source implemented with OpenESPI.

NAESB WEQ 015 (North American Energy Standards Board Wholesale Electric Quadrant)
A standard that outlines the business practices for wholesale electricity demand response programs, particularly measurement and verification. It is a part of PAP09 focused on standardized demand response signals.

NAESB WEQ19, REQ 18
This set of open standards from the North American Energy Standards Board details two-way flows of energy usage information based on a standardized model. This exchange of usage information will allow users to manage their energy consumption by providing real-time communication with utilities on power availability and cost.

NAESCO (National Association of Energy Service Companies)
A national trade association focused on energy efficiency. Members include ESCOs (Energy Service Companies), distribution companies, distributed generation companies, engineers, consultants, and finance companies. (http://www.naesco.org/)

Nameplate capacity
The maximum rated output of a generator under specific conditions designated by the manufacturer. Also known as generator nameplate capacity.

NANOG (North American Network Operators' Group)
An educational and operational group that facilitates and disseminates information related to backbone/enterprise networking technologies and

operational practices concerning the creation, maintenance, and operation of Internet Protocol networks. (http://www.nanog.org/)

NAPE (National Association of Power Engineers)
An association dedicated to education, it provides educational courses to bring new information and technology to power engineers. (http://www.powerengineers.com/)

NAPEE (National Action Plan for Energy Efficiency)
A private-public initiative begun in 2005 to create a national commitment to energy efficiency through collaboration of gas and electric utilities, utility regulators, and other partner organizations. Members include 42 utility commissions or other state/local agencies and 34 utilities. The National Action Plan identifies the key investment barriers, outlines recommendations for cost-effective energy efficiency, and offers a policy framework called Vision for 2025 to achieve and measure progress toward the goal. Both the DOE (Department of Energy) and the EPA (Environmental Protection Agency) have facilitator roles with this organization. (http://www1.eere.energy.gov/office_eere/napee.html)

Narrowband
Information capacity or bandwidth that is lower than 64 kbps.

NARUC (National Association of Regulatory Utility Commissioners)
An association representing USA and global public service commissioners who regulate electric, gas, and water utilities and other services throughout the country. It is working with FERC (Federal Energy Regulatory Commission) to develop a database of nationwide advanced metering and Smart Grid pilots. It is also coordinating with the FERC to encourage development of demand response programs that address retail and wholesale market considerations and exploring linkages between energy efficiency, demand response, and Smart Grid programs. (www.naruc.org/)

NAS (National Academy of Sciences)
One of four nonprofit, honorific societies of resources engaged in scientific and engineering research. It provides advice on science and technology issues and studies specific concerns. Most science policy and technical work is conducted by its National Research Council and volunteer committees of the nation's top scientists, engineers, and other experts. The Academy recently published a study of the USA infrastructure titled "Sustainable Critical Infrastructure Systems." It includes a review of the power infrastructure and suggests a need for a new paradigm for the renewal of all critical infrastructure systems. Membership is composed of approximately 2,100 elected members and 380 foreign associates, of whom nearly 200 have won Nobel Prizes.
(http://www.nationalacademies.org/)

NaS (Sodium-Sulfur)
Energy storage technology typically used for bulk storage by power and telecom utilities for peak shaving, backup power, firming-wind capacity, and other applications that stabilize renewable energy output. NaS battery cells are about 89 percent efficient.

NASEO (National Association of State Energy Officials)
A nonprofit organization that represents the governor-designated energy officials from each state and territory. It was created to improve the effectiveness and quality of state energy programs and policies, provide policy input and analysis, and share successes among the states.
(www.naseo.org/index.html)

NASPI (North American SynchroPhasor Initiative)
A collaborative initiative between the DOE (Department of Energy), NERC (North American Electric Reliability Corporation), and electric utilities, vendors, consultants, and researchers. It receives funding from the DOE, NERC, and industry. Its mission is to improve power system reliability and visibility through wide-area measurement and control, using the precise,

synchronized measurements of Synchrophasor technology as a diagnostic tool. Synchrophasor measurements will assist in wide-area monitoring, real-time operations, power system planning, and forensic analysis of grid disturbances. Phasor technology is expected to help integrate renewable and intermittent resources, automate controls for transmission and demand response, increase transmission system throughput, and improve system modeling and planning. The DOE has several grant programs for large-scale prototypes, regional demonstrations, and Smart Grid/PMU (Phasor Measurement Unit) deployments. (http://www.naspi.org/)

NASPInet
A conceptual architecture for an "industrial grade," secure, standardized, distributed, and expandable data communications infrastructure for synchrophasor applications in North America. Its key communications architecture elements are the phasor gateway that links PMUs (Phasor Measurement Units) to the network, the data bus that carries data between gateways, and the reference bus, a virtual bus that carries phase angle reference information based on averaging phase angles from selected reference PMUs.

National Broadband Plan
In early 2009, Congress directed the FCC (Federal Communications Commission) to develop a plan to ensure every American has "access to broadband capability." It also required a detailed strategy for achieving affordability and maximizing use of broadband to advance "consumer welfare, civic participation, public safety and homeland security, community development, health care delivery, energy independence and efficiency, education, employee training, private sector investment, entrepreneurial activity, job creation and economic growth, and other national purposes." A broadband-enabled Smart Grid could increase energy independence and efficiency. (http://www.broadband.gov/)

National grid
A concept for a future grid that provides links between today's regional grid layouts and planned renewable energy generators to transmit electricity to any region where needed. The vision for this grid is to also eliminate congestion problems and balance loads from intermittent energy sources across regions. Also known as a super grid or mega grid.

National Response Framework/Emergency Support Function 12
A USA policy document that contains the guiding principles, roles, and structures that enable response partners to prepare for and provide a unified national response to disasters and emergencies. ESF #12 describes DOE responsibility of maintaining continuous and reliable energy supplies through preventive measures and restoration and recovery actions.

National Water Information System
A database of real-time water information for the USA. It includes information on stream flows, groundwater, precipitation, lakes and reservoirs. Data is provided by the United States Geological Survey (USGS), a part of the Department of the Interior. (http://waterdata.usgs.gov/nwis/rt)

Natural water
Water within a geographic region that has developed without human intervention, in which natural processes continue to take place.

NAWC (National Association of Water Companies)
An organization that represents the private water service industry and advocates for ownership of regulated drinking water and wastewater utilities and public-private partnerships and management contract arrangements. (http://www.nawc.org/)

NBSO (New Brunswick System Operator)
An independent not-for-profit, statutory corporation established in 2004 by the New Brunswick Electricity Act. As an ISO (Independent System Operator), NBSO is primarily responsible for ensuring the reliability of the integrated electricity system and also for facilitating the operation of a competitive electricity market in New Brunswick. It is one of the 10 ISO/RTOs (Regional Transmission Organizations) currently operating in North America. (www.nbso.ca)

NCEP (National Council on Electricity Policy)
A group of policy and industry associations focused on improving coordination of electricity policy between federal and state organizations. It consists of NARUC (National Association of Regulatory Utility Commissions), NASEO (National Association of State Energy Officials), the National Conference of State Legislatures, the National Association of Clean Air Agencies, the National Governors Association, FERC (Federal Energy Regulatory Commission), DOE (Department of Energy), and EPA (Environmental Protection Agency). (http://www.ncouncil.org/)

NCOIC (Network Centric Operations Industry Consortium)
An international organization of governments and industry promoting network and systems interoperability based on open standards, processes, and architectures for secure and reliable communications. Their primary focus is on military and air traffic management applications.

NDFD (National Digital Forecast Database)
One of the foundations of the National Weather Service's digital services program, NDFD collects data and forecasts via its field stations on atmospheric activity, including cloud cover and temperature data. It uses XML to return responses in a language called DWML (Digital Weather Markup Language).

NDT (Non-Destructive Testing)
A term used in inline pipeline inspections to describe techniques and technologies that can monitor or test pipeline integrity without destroying that infrastructure.

NDWC (National Drinking Water Clearinghouse)
An organization sponsored by the Rural Utilities Service (RUS) dedicated to helping small communities through information about sustainable development and drinking water issues. (http://www.nesc.wvu.edu/drinkingwater.cfm)

NEB (National Energy Board)
An independent federal agency that regulates several aspects of Canada's energy industry. It promotes safety and security, environmental protection, and efficient energy infrastructure and markets in Canada according to the mandate set by Parliament regarding pipelines, energy development, and trade. (www.neb.gc.ca)

NEC (National Electrical Code®)
A set of regulations to ensure that safe electrical systems are designed and installed in the USA. The NEC is updated every three years, and its regulations are generally accepted across the country. Sponsored by the NFPA (National Fire Protection Association), NEC is also known as NFPA 70. Its next update will occur in 2014. (http://www.nfpa.org/aboutthecodes/AboutTheCodes.asp?DocNum=70)

NEC Article 393
An article of the NEC code about low voltage suspended ceiling power distribution systems. It addresses equipment connected to indoor ceiling grids, floors, and walls. Also included are power supply and wiring requirements for DC lighting. It enables building level microgrids and PV powered lighting.

NEC 625
An article of the NEC code that defines regulations for EV charging systems, including wiring methods, equipment construction, controls, protection, and supply equipment locations.

NECF (National Energy Customer Framework)
An Australian program focused on reforming and consolidating various state and territory energy distribution and retail regulations into one national set of laws and rules.
(http://www.ret.gov.au/energy/energy_markets/national_energy_custome r_framework/Pages/NationalEnergyCustomerFramework.aspx)

NEDO (New Energy and Industrial Technology Development Organization)
An organization established by the Japanese government in 1980 to develop new oil-alternative energy technologies. Since then it has added industrial and environmental technology research and development, and supports the JSCA (Japan Smart Community Alliance).
(http://www.nedo.go.jp/english/introducing/what.html)

NEEA (Northwest Energy Efficiency Alliance)
A nonprofit organization focused on development and adoption of energy-efficient products and services. It is supported by the region's electric utilities, state governments, public interest groups, and industry representatives. It operates programs in Idaho, Montana, Oregon, and Washington. It is funded by leading Northwest electric utilities, the Energy Trust of Oregon, and the Bonneville Power Administration.
(http://www.nwalliance.org/)

NEED (National Energy Education Development Project)
A nonprofit educational association whose mission is to promote an energy-conscious society. The NEED Project creates networks of students, educators, business, government, and community leaders to design and deliver energy education programs for schools. NEED partners with the

DOE's (Department of Energy's) EIA (Energy Information Administration) to obtain the data and energy analysis used to annually update NEED teaching materials. (http://www.need.org/)

NEEP (Northeast Energy Efficiency Partners)
A nonprofit organization founded in 1996 whose mission is to promote the efficient use of energy in homes, buildings, and industry in the Northeast United States. (http://www.neep.org)

NEETRAC (National Electric Energy Testing & Research Applications Center)
A nonprofit electric energy research, development, and testing center within the Georgia Institute of Technology's School of Electrical and Computer Engineering. It is involved in industry-sponsored research to improve the quality of transmission and distribution systems. (http://www.neetrac.gatech.edu/)

Negawatt
Watts of energy reduced on a temporary basis in response to a market signal – usually price. It is the outcome of Demand Response programs that aggregate a number of these actions to represent reductions of energy use from kilowatts to megawatts. A permanent negawatt reduction is achieved through energy efficiency programs and actions.

NEMA (National Electrical Manufacturers Association)
A USA trade association of about 450 companies that manufacture products used in the generation, transmission, distribution, control, and end use of electricity. It helps to develop and promote the IEC's standards in the USA. (www.nema.org)

NEMA SG-AMI 1-2009
Guidelines to help determine compliance to security and functionality requirements for smart meter firmware upgradeability in AMI networks.

NEMS (The National Energy Modeling System)
A computer-based, energy-economy modeling system of USA energy markets that has a 25-year forward view. It is used by the EIA to project the energy, economic, environmental, and security impacts on the USA based on different assumptions about energy markets. The Electricity Market Module represents generation, transmission, and pricing of electricity, subject to delivered prices for coal, petroleum products, natural gas, and biofuels; costs of generation by all generation plants, including capital costs and macroeconomic variables for costs of capital and domestic investment; environmental emissions laws and regulations; and electricity load shapes and demand. There are three primary submodules: capacity planning, fuel dispatching, and finance and pricing. (http://www.eia.doe.gov/oiaf/aeo/assumption/introduction.html)

NEPA (National Environmental Policy Act)
A federal act that requires federal agencies to integrate environmental values into their decision-making processes by considering the environmental impacts of their proposed actions and reasonable alternatives to those actions. (http://www.epa.gov/Compliance/nepa/)

NEPOOL (New England Power Pool)
A voluntary association formed in 1971 by the New England region's private and municipal utilities to foster cooperation and coordination among utilities in the six-state region. NEPOOL created a regional power grid that now includes more than 300 separate generating plants and more than 8,000 miles of transmission lines—all interconnected and dedicated to ensuring that New England never again has a region-wide power failure. As part of FERC's (Federal Energy Regulatory Commission's) restructuring of wholesale electric power, it is now a group of generators, utilities, marketers, public power companies, and end users within Maine, New Hampshire, Vermont, Massachusetts, Connecticut, and Rhode Island. It is part of the ISO-NE (Independent System Operator – New England). (http://www.iso-ne.com/)

NERC (North American Electric Reliability Corporation)
An international, independent, self-regulatory, not-for-profit organization whose mission is to ensure the reliability of the bulk power system in North America. It monitors the bulk power system; develops and enforces reliability standards; assesses future adequacy of electricity; audits owners, operators, and users for preparedness; and educates and trains industry personnel. The standards that it proposes are reviewed by FERC (Federal Energy Regulatory Commission) and Canadian regulators for adoption. NERC works with eight Regional Entities to conduct its mission (FRCC, MRO, NPCC, RFC, SERC, SPP, TRE, and WECC). It sets standards for grid reliability and cyber security, working with the FERC on the Critical Cyber Asset requirements. (http://www.nerc.com/)

NERR (National Energy Retail Rules)
A harmonized national consumer protection framework for the retail sale of electricity and gas that is administered by the AEMC (Australian Energy Market Commission). (http://www.mce.gov.au/emr/rpwg/default.html)

NESCO (National Electric Sector Cybersecurity Organization)
An organization of governmental and industry stakeholders that is funded by DOE (Department of Energy) and focuses on strengthening electrical sector cybersecurity through public-private partnerships. (http://energy.gov/oe/services/cybersecurity/nesco)

NESCOR (National Electric Sector Cybersecurity Organization Resource)
The research and analysis team for NESCO that is staffed by EPRI (Electric Power Research Institute).

Net energy to load
A NERC (North American Electric Reliability Corporation) formula used to help standardize data collection from its members for demand response programs. This calculation is the amount of net Balancing Authority (BA) area generation, plus energy received from other Balancing Authority

areas, less energy delivered to Balancing Authority areas through interchange. It includes Balancing Authority area losses but excludes energy required for storage at energy storage facilities.

Net excess power
As defined in the ARRA Section 1306 amending EISA 2007 (Energy Independence and Security Act of 2007), any quantity of electricity that is facility-generated, recovered, or stored that exceeds the consumption needs of the facility.

Net internal demand
A NERC (North American Electric Reliability Corporation) formula used to help standardize data collection from its members for demand response programs. This calculation is the total internal demand reduced by direct control load management and interruptible demand.

Net metering
The capability for residential and C&I (Commercial and Industrial) customers to generate electricity and sell back excess power to the utility, essentially offsetting their future purchases of utility power. Net metering uses either a single, bi-directional electric meter or two meters to separately measure in and out electricity flows at a customer's location. Net metering is currently implemented on a state-by-state basis with significant variation between states.

Net savings
Energy efficiency program savings as one of three levels. Net savings means gross savings adjusted for market effects like free-riders. The other two measures are ex ante savings and gross savings.

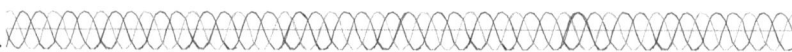

NETCONF Configuration Protocol
This standard developed by the IETF (Internet Engineering Task Force) provides mechanisms to install, manipulate, and delete the configuration of networked devices in an XML-based data configuration. In June 2011 it was republished as RFC 6241.

NETL (National Energy Technology Laboratory)
A DOE (Department of Energy) lab committed to enhancing America's energy security, improving the environmental acceptability of energy production and use, increasing the competitiveness and reliability of USA energy systems, and ensuring a robust energy future. Unlike other DOE laboratories, this one is federally owned and operated and devotes the majority of its funding to R&D partnerships with industry, university, and other government entities. It manages the Modern Grid Strategy and T&D (Transmission and Distribution) research and development projects for grid modernization and assesses the benefits of distributed generation. (http://www.netl.doe.gov/index.html)

NETMAN (Network Management Initiative)
This is an integrated set of standards for management of physical, virtual, and software defined networks. The purpose of this initiative is to work towards the unification of network management practices across datacenters, cloud infrastructure, and Network Function Virtualization applications.

Network Intelligence Alliance
An industry organization comprised of companies that capture, process, decode, or analyze network data. The focus is on collaborative knowledge development and market education about the roles of network intelligence. http://www.nialliance.org/members/

Network-of-networks
A network comprised of smaller, heterogeneous public and private networks that connect to each other. It is commonly used in association with the Internet, but will be equally applicable to a fully enabled Smart Grid.

New York Energy Highway
A public-private initiative to upgrade and modernize New York State's electric power system with significant additions of renewable energy and energy efficiency solutions. (http://www.nyenergyhighway.com/)

NFC (Near Field Communications)
A form of M2M communications in which tags and devices wirelessly communicate when in close proximity (touch or up to 4 centimeters) to other similarly-enabled devices.

NFC Forum
An industry association of vendors and end users that promotes the use of NFC technology in consumer electronics, mobile devices, and PCs through standard specifications and a certification program. (http://www.nfc-forum.org)

NFC tag (Near Field Communication tag)
A passive device that stores data that can be read by an NFC-enabled device. It typically broadcasts the same data to all NFC devices and does not exchange data with them.

NFCRC (National Fuel Cell Research Center)
A research facility at the University of California-Irvine established in 1998 by DOE (Department of Energy) and CEC (California Energy Commission) to accelerate the development and deployment of fuel cell technology. It is principally focused on stationary power and its role as a Distributed

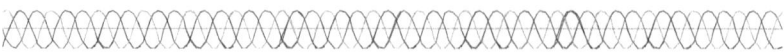

Generation and Central Power Plant technology.
(http://www.nfcrc.uci.edu/2/default.aspx)

NFPA (National Fire Protection Association)
An international nonprofit organization focused on advocating codes and standards, education, training, and research to improve fire safety and reduce hazards. A number of standards have application to the electrical grid. (http://www.nfpa.org/)

NFPA 70 (National Fire Protection Association 70)
A set of regulations to ensure that safe electrical systems are designed and installed in the USA. It is updated every three years, and its regulations are generally accepted across the country. Sponsored by the NFPA (National Fire Protection Association), NFPA 70 is also known as the National Electric Code.

NFPA 70E (National Fire Protection Association 70E)
A standard that offers guidance to reduce hazards involved in the installation, inspection, operation, maintenance, and demolition of electric conductors, electric equipment, signaling and communications conductors and equipment, and raceways. It also includes safe work practices.

NFPA 110 (National Fire Protection Association 110)
A standard that covers installation, maintenance, operation, testing and performance requirements for emergency and standby power systems that are alternate sources of electrical power when primary power sources fail.

NFPA 111 (National Fire Protection Association 111)
A standard that addresses installation, maintenance, operation, testing, and performance requirements for stored electrical energy systems that are alternate sources of electrical power in buildings and facilities when primary power sources fail.

NGR (NonGenerator Resource)
A CAISO term for resources like demand response that could provide ancillary services for grid stability.

NGWA (The National Ground Water Association)
An international nonprofit organization focused on scientific and economic information about groundwater and its development, protection and management. The organization also conducts advocacy programs on behalf of its members. (http://www.ngwa.org/)

NiCd (Nickel Cadmium)
A rechargeable battery technology with limited use in power systems. It has limited recharge capability and uses very toxic materials.

NIETC (National Interest Electric Transmission Corridors)
The subject of a study conducted by the DOE's (Department of Energy) Office of Electricity Delivery and Energy Reliability. It identifies two transmission corridors: one in the Mid-Atlantic USA and the other in the Southwest USA (http://nietc.anl.gov/)

NiMH (Nickel Metal Hydride)
NiMH batteries are similar in properties to NiCd batteries, but are considered to have a less harmful environmental impact than NiCd batteries due to absence of toxic cadmium.

NIPP (National Infrastructure Protection Plan)
A program within DHS (Department of Homeland Security) targeted to create a national plan that delivers enhanced protection and resiliency of the nation's critical infrastructure and key resources. Both electric systems and communication systems are identified as part of the critical infrastructure. DOE (Department of Energy) is the agency working with DHS to identify a vision of a robust, resilient energy infrastructure that maintains

service and business continuity and build a framework to achieve that vision. (http://www.dhs.gov/xprevprot/programs/editorial_0827.shtm)

NIPPC (Northwest & Intermountain Power Producers Coalition) A group of public and private power producers in the Pacific Northwest and Mountain regions, whose mission is to ensure a competitive electric power supply marketplace. These methods include favorable laws, policies, rules and regulations, with an emphasis on independently produced and/or renewable power sources. (www. Nippc.org)

NIST (National Institute of Standards and Technology) Part of the USA DOC (Department of Commerce), NIST is responsible for development of a Smart Grid architecture, interoperability, cyber security, and implementation standards for the Smart Grid, and development of testing and certification frameworks for equipment and systems conformance to these standards. NIST received $10 million in funding in Title VI, Section 1305 of ARRA (2009 Stimulus Bill) and is tasked to work with federal, state, and local agencies and private industry consortiums, such as the Gridwise Architecture Council, NEMA (National Electrical Manufacturers Association), IEEE (Institute of Electrical and Electronics Engineers), EPRI (Electric Power Research Institute), and NERC (North American Electric Reliability Corporation) to create these standards and frameworks. Eight SGIP (Smart Grid Interoperability Panel) Working Groups are focused on TnD (Transmission and Distribution); B2G (Building to Grid) for commercial building integration to the grid; H2G (Home to Grid) for residential; I2G (Industrial to Grid); V2G (Vehicle to Grid); B&P (Business and Policy), which helps connect the technical work to higher-level business and political issues; Terminology; and Electromagnetic Interoperability Issues. The standards recommendations will be submitted for review and approval by FERC (Federal Energy Regulatory Commission). (www.nist.gov/smartgrid/)

NIST 800-39
The Guide for Applying the Risk Management Framework to Federal
Information systems describes near real-time risk management via a
monitoring process.

NIST Smart Grid Framework 1.0
Phase 1 of a three-phase NIST (National Institute of Standards and
Technology) plan to accelerate the identification of standards while
establishing a robust framework for the longer-term evolution of the
standards and establishment of testing and certification procedures. It
describes a high-level reference model for the Smart Grid, identifies nearly
80 existing standards that can be used now to support Smart Grid
development, identifies 15 high-priority gaps, plus cyber security, for which
new or revised standards are needed, documents action plans with
deadlines for designated SDOs to fill these gaps, and describes the strategy
being pursued to establish standards for ensuring cyber security of the
Smart Grid.

NIST SP-500- 267
This is a profile for IPv6 (Internet Protocol version 6) IPv6 for use by the U.S.
Government's IT systems and applications. It outlines a set of common
network devices and minimum requirements for IPv6 compatibility.

NIST SP-800-53 Rev.3
One of two special publications from NIST (National Institute of Standards
and Technology) that offers guidelines for secure interoperability for
federal agencies who are part of the bulk power system that may influence
Smart Grid security and interoperability definitions. It is titled *DRAFT
Recommended Security Controls for Federal Information Systems and
Organizations*. (http://csrc.nist.gov/publications/PubsDrafts.html)

NIST SP-800-82
One of two special publications from NIST (National Institute of Standards and Technology) with guidelines for secure interoperability for systems including SCADA (Supervisory Control And Data Acquisition). It identifies typical threats and vulnerabilities to these systems, and offers recommendations on risk reduction actions. It is titled *DRAFT Guide to Industrial Control Systems (ICS) Security*.
(http://csrc.nist.gov/publications/PubsDrafts.html)

NISTIR (National Institute of Standards and Technology Interagency Report) A NIST report that describes research to a specialized audience. Also known as a NIST Internal Report.

NISTIR 7628
A NIST (National Institute of Standards and Technology) document that provides Guidelines for Smart Grid Cyber Security, comprised of an introduction and 3 volumes. It is produced by the Smart Grid Interoperability Panel Cyber Security Working Group and has been adopted by the U.S. Department of Homeland Security for government cyber security applications.
(http://csrc.nist.gov/publications/nistir/ir7628/introduction-to-nistir-7628.pdf)

NISTIR 7628 Volume 2 Revision 1: Privacy and the Smart Grid
This document was completed by the NIST-led Smart Grid Cybersecurity Committee (formerly the Cyber Security Working Group) of the Smart Grid Interoperability Panel. The updated document addresses changes in technologies, implementations, and new privacy risks since the release of NISTIR 7628 in September 2010. Use cases were included, training and awareness tools were created and included, and a section discussing legal issues within the smart grid was included. This version was published in September 2014.

NISTIR 7761
This is a set of tools and methods to guide and advise Smart Grid communications system designers in making decisions regarding existing and emerging wireless technologies for potential use in AMI (Advanced Metering Infrastructure) and DA (Distribution Automation).

NIWR (The National Institutes for Water Resources)
Research entities housed at the land grant universities in each state, the District of Columbia, the Virgin Islands, Puerto Rico and Guam that communicate research needs from states and localities and assess research priorities. Authorized by the Water Resources Research Act of 1964, these entities work to solve water supply and quality problems at the local, state, regional and national levels. (http://snr.unl.edu/niwr/)

NLoS (non-line of sight)
Frequencies and signaling technologies that do not require transmitters and receiver antennas to be within unobstructed sight of each other.

NMS (Network Management System)
Software and hardware that gives utilities the tools to manage the energy distribution grid in normal, outage, and emergency conditions. Normal conditions include load balance, regulatory and operational reporting, creation of work orders, and collection of crew reports. Outage conditions can trigger information exchanges with customer communications, load switching, and work crew dispatch applications. Emergency conditions activities include damage assessment and collection and analysis of management data.

NoC (Network-on-Chip)
A communication subsystem on an integrated circuit that exists between heterogeneous IP cores and blocks on SoCs (System on Chips). It creates greater scalability for SoCs, routes data more efficiently, reduces congestion and increases system speed.

Non-dispatchable demand response
Voluntary (on the part of the consumer) demand response reductions based on pricing structures like TOU, CPP, real-time pricing, and system peak response transmission tariffs and are not based on instruction from a control center.

Non-returnable storage
Conversion of electricity into another useable form of energy instead of returned to the grid as electricity. This stored energy can be used in CAES, ice, or thermal storage.

Non-spinning reserve
From a demand response perspective, generating reserve that is offline or not connected to the system but capable of serving demand within a specified time. It can also include interruptible load that can be removed from the system in a specified time. Specified times may vary by ISO or region.

Nonce
An abbreviation for "number used once" that is commonly used in cryptography to a secure conversation by using a random number to begin to authenticate communications with another party.

Nonconsumptive water use
Water use that does not reduce supply – it includes uses such as recreation and some power production.

Nonpotable water
Water that is not fit for human consumption.

NOPE (Not On Planet Earth)
An even more extreme example of BANANA (Build Absolutely Nothing Anywhere Near Anything) and NIMBY (Not In My Backyard).

NOPR (Notice of Proposed Rule-making)
A FERC (Federal Energy Regulatory Commission) term and step in its rule-making process. The NOPR is the second step of the process in which the recommended proposal or rule is available for public comment and then issued after staff review and revisions into an Order.

NordREG (Nordic Energy Regulators)
An organization for Nordic energy regulators with the mission to promote legal and institutional frameworks and conditions necessary for development of Nordic and European electricity markets. (https://www.nordicenergyregulators.org/)

NPCC (Northeast Power Coordinating Council)
A council that promotes and enhances the reliable and efficient operation of the international, interconnected bulk power system in Northeastern North America through development and enforcement of regional reliability standards, coordination of system planning, design and operations, and assessment of reliability. NPCC is one of the eight NERC (North America Electric Reliability Corporation) Regional Entities or reliability regions. The geographic area covered by NPCC includes New York, Maine, Vermont, New Hampshire, Connecticut, Massachusetts, Rhode Island, and Ontario, Québec, and the Maritime Provinces. (www.npcc.org/aboutus/about.aspx)

NPCC (Northwest Power and Conservation Council)
Created by Congress in the Northwest Power Act of 1980 and funded by wholesale power revenues from the Bonneville Power Administration, this council creates a 20-year plan to ensure adequate and reliable energy at the lowest economic and environmental cost to the Northwest. Energy conservation and renewable resources are priorities identified in the Power Act. (http://www.nwcouncil.org/)

NPDES (National Pollutant Discharge Elimination System)
An EPA permit program that controls water pollution by regulating point sources that discharge pollutants into US waters. (http://cfpub.epa.gov/npdes/)

NPLC (Number of Power Line Cycles)
The number of power line cycles in one second. This measurement can also be displayed in Hertz, and is an important statistic for measuring power behavior over a period of time. It is used for DC voltage and current, two and four wire resistance as well as temperature measurements.

NRA (National Regulatory Authority)
A European acronym for European Commission member states' in-country regulatory agencies for electricity and gas.

NRECA (National Rural Electric Cooperative Association)
An industry group that represents 864 distribution and 66 generation and transmission cooperatives serving over 40 million people in 47 states, approximately 12 percent of the USA population. These utilities own 42 percent of the electric distribution lines and deliver 10 percent of the total kWh sold in the USA each year. (http://www.nreca.org/)

NREL (National Renewable Energy Laboratory)
A DOE (Department of Energy) national laboratory, it develops renewable energy and energy efficiency technologies and practices, advances related science and engineering, and transfers knowledge and innovations to address the nation's energy and environmental goals. Many of NREL's major research programs have analysis functions—from life-cycle to vehicle systems to online renewable energy analysis applications and modeling tools. Most of these tools can be applied on a global, regional, local, or project basis. HOMER (Hybrid Optimization Model for Electric Renewables) is one of NREL's tools. (http://www.nrel.gov/)

NRF (National Response Framework)
Guiding principles to prepare for and provide a unified national response for small to large disasters and emergencies. It defines the key principles, roles, and structures for organized response from communities, tribes, States, the Federal Government, and private-sector and nongovernmental partners.

NRF/ESF 12 (National Response Framework/Emergency Support Function 12)
ESFs are mechanisms to coordinate functional capabilities and resources provided by federal USA departments and agencies, along with certain private-sector and nongovernmental organizations under the NRF. ESF 12 addresses energy, and the agencies activated by an incident collect, evaluate, and share information on energy system damage and estimations on the impact of energy system outages within affected areas. The collective agencies monitor energy restoration processes and facilitate the restoration of energy systems through legal authorities and waivers.

NRRI (National Regulatory Research Institute)
A nonprofit founded in 1976 by NARUC (National Association of Regulatory Utility Commissioners) as a resource for state public utility commissions. It is focused on identifying regulatory challenges and creating new knowledge and democratizing existing knowledge to empower utility regulators to make public interest decisions of the highest possible quality. (http://nrri.org/)

NRWA (National Rural Water Association)
A water and wastewater utility organization representing over 25,000 public water and wastewater utilities. It delivers education, technical expertise, and policy and regulatory advocacy about rural water system issues. (http://www.nrwa.org/)

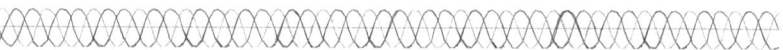

NSF
An independent, not-for-profit, non-governmental organization that works with the American National Standards Institute (ANSI) to develop public health and safety standards. It conducts global testing and certification programs for drinking water treatment units, drinking water treatment chemicals and drinking water system components. NSF originally meant National Sanitation Foundation, but the organization is now only known by the acronym. (http://www.nsf.org/business/about_NSF/)

NSF/ANSI 58-2006: Reverse Osmosis Drinking Water Treatment Systems
A standard that addresses point-of-use reverse osmosis drinking water treatment systems to reduce total dissolved solids and other specific contaminants.

NSF/ANSI Standard 60: Drinking Water Treatment Chemicals
A health effects standard for chemicals that are used to treat drinking water. Chemicals created as reaction by-products of the process are not covered.

NSF/ANSI Standard 61: Drinking Water System Components
A health effects standard for all devices, components and materials which contact drinking water. It establishes the minimum health effects required for anything (components, materials, products or systems) that contact the drinking water and/or chemicals used in processing or treatment. It does not cover items affecting performance criteria of drinking water systems, such as taste or safety aspects.

NSTB (National SCADA Test Bed)
A national test facility to help secure SCADA (Supervisory Control And Data Acquisition) communications and controls within the energy sector. It combines the expertise and resources of several national labs into a multi-lab partnership that helps to identify and correct critical security flaws in control systems and equipment. Established by DOE's (Department of

Energy) OE (Office of Electricity Delivery and Energy Reliability), the test bed offers a full-scale infrastructure suite of facilities for testing and validating control systems. The NSTB offers the integrated expertise and resources of multiple national labs, including Idaho National Laboratory, Sandia National Laboratories, Argonne National Laboratory, Pacific Northwest National Laboratory, and Oak Ridge National Laboratory. (http://www.oe.energy.gov/nstb.htm)

NTA 8150 (Netherlands Technical Agreement 8150)
A protocol for interoperable Smart Grid communications between different web-based and enterprise software applications running within utilities promoted by ESNA (Energy Services Network Association) and Echelon.

NTCIP 1213 (National Transportation Communications for ITS Protocol 1213)
This is a standard created by the US Department of Transportation's ITS (Intelligent Transportation System). The standard addresses the protocol and points of interoperability for remote monitoring and control of street and highway electrical assets such as roadway lighting.

NTSB (National Transportation Safety Board)
An independent investigatory agency that does not have regulatory oversight. The Pipeline and Hazardous Materials Division is responsible for investigation of natural gas transmission accidents. (http://www.ntsb.gov/)

NTIA (National Telecommunications and Information Administration)
A DOC agency that is responsible for advising the President on telecommunications and information policies. It manages Federal spectrum and is involved in the national broadband programs. (http://www.ntia.doc.gov/)

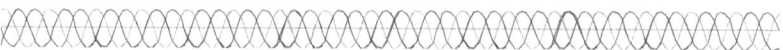

NTTRC (National Transmission Technology Research Center)
Part of ORNL, it is a test and evaluation facility for critical transmission systems and technologies. It includes the Outdoor PCAT (Powerline Conductor Accelerated Testing) facility and is planning an indoor PCAT facility, a powerline conductor operational testing facility, and a transmission power electronics test bed. (http://www.ornl.gov/sci/oetd/facilities.htm)

NUG (Non-Utility Generator)
An entity that owns or operates electric generation facilities to sell power to the public that is not an electric utility. Another term is IPP (Independent Power Producer).

NWP (Numerical Weather Prediction)
Weather forecasting accomplished with atmospheric models that use equations based on physics principles. This type of forecasting is used for wind power production forecasts, usually in combination with statistical models.

NWPP (Northwest Power Pool)
A voluntary organization that comprises major generating utilities serving the Northwestern USA, British Columbia, and Alberta. Smaller, principally non-generating utilities in the region participate indirectly. NWPP promotes cooperation among its members to achieve reliable operation of the electrical power system, coordinate power system planning, and assist in transmission planning in the Northwest Interconnected Area. It covers Washington, Oregon, Idaho, Utah, and parts of Montana, Wyoming, and Nevada. (http://www.nwpp.org/index.html)

NWRA (National Water Resources Association)
A nonprofit federation of rural water districts, municipal water entities, businesses and individuals that is focused on policy and regulation for the

management, conservation and use of water and land resources across the nation. (http://www.nwra.org/)

NYBEST^{TM} (New York Battery and Energy Storage Technology) **Consortium**
An industry organization focused on building positioning New York State as a global leader in energy storage technology through commercial introductions of transportation, grid storage, and power electronics solutions with funding from NYSERDA. Members include vendors, universities, and utilities. (http://www.ny-best.org/)

NYISO (New York Independent System Operator)
A nonprofit ISO organization established in 1999 that operates New York's bulk electricity grid and wholesale electricity markets and provides comprehensive reliability planning for the state's bulk electricity system. It covers the state of New York and is one of the 10 ISO/RTOs (Regional Transmission Organizations) currently operating in North America. (http://www.nyiso.com/public/index.jsp)

NYSERDA (New York State Energy Research and Development Authority)
A public benefit corporation created in 1975 under Article 8, Title 9 of the New York State Public Authorities Law, primarily funded by state rate payers. It is governed by a board that includes commissioners of the Department of Transportation, Department of Environmental Conservation, Public Service Commission, and state's Power Authority. NYSERDA focuses on developing a diversified energy supply portfolio, improving market mechanisms, and promoting the adoption of advanced technologies to plan for and respond to uncertainties in the energy markets. It recently released Program Opportunity Notice, PON-1200, providing up to $1.5 million in matching funds for grid-connected energy storage demonstration projects in New York State. (http://www.nyserda.org/)

NYSRC (New York State Reliability Council)
A not-for-profit entity with the mission to promote and preserve the reliability of electric service in New York state by developing, maintaining, and, from time to time, updating the Reliability Rules governing NYISO and all entities engaging in electric transmission, ancillary services, energy, and power transactions on the state power system. (http://www.nysrc.org/)

NYSSC (New York State SmartGrid Consortium)
A nonprofit public-private organization that promotes New York's implementation of a safe, secure, and reliable smart grid. Members include utilities, technology vendors, research institutions, government, and quasi-government entities. (http://nyssmartgrid.com)

O

O & M (Operations and Maintenance)
A shorthand phrase for two important utility functions focused on physical plant facilities to ensure optimal performance.

OAIF (Office of Integrated Analysis and Forecasting)
An office located in the DOE's (Department of Energy) EIA (Energy Information Administration), it conducts forward-looking analyses and develops mid- to long-term projections of alternative energy futures for both domestic and international energy markets. It also measures and estimates greenhouse gas emissions and manages the voluntary program for reporting greenhouse gas. The Office publishes the Annual Energy Outlook, International Energy Outlook, Emissions of Greenhouse Gases in the United States, and Voluntary Report of Greenhouse Gas Emissions, along with other analyses as requested by the government. (http://www.eia.doe.gov/oiaf/aeo/index.html)

OASIS (Open Access Same-time Information System)
An Internet-based tool for sharing information on transmission prices and product availability. It is the primary means by which high-voltage transmission lines are reserved for moving wholesale quantities of electricity. FERC (Federal Energy Regulatory Commission) restricts communication between power marketing and transmission operation employees within any one organization. Utilities can obtain information about their own transmission system for their own wholesale power transactions only through OASIS. OASIS was conceptualized with the Energy Policy Act of 1992 and formalized in 1996 through FERC Orders 888 and 889. This was previously known as TSIN (Transmission System Information Networks).

OASIS (Organization for the Advancement of Structured Information Standards)
A global not-for-profit organization focused on the development, convergence, and adoption of standards for electronic commerce and Web services. It is collaborating with DRRC (Demand Response Research Center) on the OpenADR (Open Automatic Demand Response) specification through its Energy Interoperation Technical Committee. It also works on EMIX (Energy Market Information eXchange). (http://www.oasis-open.org/)

OASIS Energy Interoperation TC
A technical committee (TC) that works to define interaction between Smart Grids and their end nodes, including Smart Buildings, Enterprises, Industry, Homes, and Vehicles. The TC develops data and communication models that enable the interoperable and standard exchange of signals for dynamic pricing, reliability, and emergencies. The TC's agenda also extends to the communication of market participation data (such as bids), load predictability, and generation information. Energy Interoperation relies on the OASIS EMIX Specification for communication of price and product defintion. The final published version of Energy Market Information Exchange (EMIX) Version 1.0 can be found in the OASIS archives; the final published version of WS-Calendar 1.0 is also in the archives. (http://www.oasis-open.org/committees/tc_home.php?wg_abbrev=energyinterop)

OASIS oBIX (Open Building Information eXchange) V1.1
An M2M communication model made by OASIS that provides the framework for the core information model and interactions with building control systems via an XML schema. This allows enterprise applications to communicate with mechanical or electrical control systems in buildings. Version 1.1 is an update of the 2006 oBIX specification that allows greater compatibility with already installed oBIX systems.

OASIS WS-Security
This is a suite of cyber security tools and standards that is currently in development to update version 1.0. It delivers a technical foundation for interoperable ecommerce web service security functions. It supports a variety of languages, protocols, and security models that provide web services confidentiality and integrity.

Oatts (Open Access Transmission Tariffs)
A regulatory mandate to allow use by others of a utility's transmission and distribution facilities to move bulk power from one point to another on a non-discriminatory fee basis.

OCCP (Open Charge Point Protocol)
A protocol supporting uniform communications between charging stations and management systems.

ODVA (Open DeviceNet Vendors Association)
An international association of industrial automation businesses promoting the Common Industrial Protocol (CIP) as a core technology for open and interoperable information exchanges and communications. (http://www.odva.org/)

OE (Office of Electricity Delivery and Energy Reliability)
The DOE (Department of Energy) office with responsibility to lead national efforts to modernize the electric grid, enhance security and reliability of the energy infrastructure, and facilitate recovery from disruptions to energy supply. As part of these duties, it runs the National SCADA Test Bed. It also oversees NETL (National Energy Technology Laboratory). It is occasionally referred to as OEDER. (http://www.oe.energy.gov/)

OEA (Office of Energy Assurance)
An office in the DOE (Department of Energy) that publishes a daily summary of public information concerning current energy issues to inform

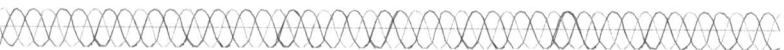

stakeholders of developments affecting energy systems (including electricity), flows, and markets.

OECD (Organization for Economic Co-operation and Development)
A global entity that promotes policies to improve the economic and social qualities for citizens through fair market economies backed by democratic governmental institutions. From Smart Grid and M2M perspectives, it has been influential in recommendations regarding data privacy and telecommunications policy. (www.oecd.org)

OFDM (Orthogonal Frequency Division Multiplexing)
A digital encoding and modulation technology used for WiMAX fixed-service 802.16d. It achieves a data throughput of more than 1 Mbps downstream through several overlapping carrier signals instead of just one. It is a signaling technology that delivers NLoS functionality. Future 4G may be based on OFDM technology.

OFDMA (Orthogonal Frequency Division Multiple Access)
The multiuser version of OFDM used in Mobile WiMAX technology and is the IEEE 802.16e-2005 standard for mobile service. OFDMA assigns subsets of sub-carriers to individual users, allowing simultaneous low data-rate transmission from several users.

Off peak period
A time frame of low demand for electricity, usually predictable by the time of day or night, day of the week, and season of the year. Off-peak periods may be at night or during the winter in a warm climate.

OGC (Open Geospatial Consortium)
A nonprofit and global organization focused on development of standards for geospatial and location-based services. The OpenGIS® standards are developed and promoted by this organization. (http://www.opengeospatial.org/)

OGC GML (Open Geospatial Consortium Geography Markup Language)
This is an open standard for the transport and storage format of geographic information transactions via the internet. Its objective is to meet the needs of geographic location based Smart Grid applications.

Ohm
A measurement that quantifies a material's resistance or impedance to electrical current.

Ohm's law
The mathematical relationship between voltage, current, and resistance was first described by Georg Ohm and it is still used to analyze electrical circuits.

OMS (Outage Management Systems)
Application software that performs grid service outage detection, analysis, diagnosis and restoration, and communication with other systems, like SCADA (Supervisory Control And Data Acquisition), AMR (Automated Meter Reading), and GIS (Geographic Information System), to transfer actionable information about changes to grid status.

On-peak period
A time frame of high demand for electricity, usually predictable by the time of day or night, day of the week, and season of the year. On-peak periods may be during the hottest months or on the hottest days.

Once-through cooling system
Part of a thermoelectric generation plant, a cooling system that withdraws water from a river, lake or ocean and discharges it back (at a higher temperature) into that same body of water.

OPC (OLE for Process Control)
A set of standard OLE/COM (component object model) interface protocols to provide interoperability between automation and control applications, field systems and devices, and enterprise applications in the process control industry. Originally based on Microsoft's OLE COM and DCOM (distributed component object model), the standards now comprise hundreds of OPC data access servers and clients.

OPC Foundation (OLE for Process Control Foundation)
A global, nonprofit organization of manufacturers and vendors in factory and process-automation equipment and solutions with a focus on developing industry standards for interoperable data transfer in manufacturing and process industries. Data may move vertically from the factory floor through the enterprise of multi-vendor systems and between devices on different industrial networks from different vendors. (http://www.opcfoundation.org/)

OPC-UA Industrial
This is the next generation industrial automation standard created by the OPC Foundation that provides a cross platform framework for systems using Java, C++, .NET, and others for access to real-time and historical data and events

OPC-UA ISA95 -/- ANSI/ISA 95 -/- ISO/IEC 62264
This standard defines models and terminology governing the integration of industrial control and business enterprise systems. It serves to bridge the occasional access requirements of the corporate sector and the high speed, fast response times required by industrial controls systems.

Open Charge Alliance
A global consortium of public and private EV (electric vehicle) infrastructure advocates to promote open standards and specifications for EVs, with

emphasis on expanding the development and adoption of OCCP (Open Charge Point Protocol). (www.openchargealliance.org)

Open grid
A grid of any size that allows a bi-directional flow of electricity.

Open Group
A not-for-profit industry consortium that promotes information accessibility within and between organizations based on open standards and interoperability. (http://www.opengroup.org/)

OPEN (Open Public Extended Network) **meter**
An EU-sponsored project to create open and public standards for AMI for electricity, gas, water and heat metering. It consists of seven work package structures and deliverables within each work package that outline requirements and testing for proposed standards for AMI and smart meters. (http://www.openmeter.com/)

OpenADE (Open Automatic Data Exchange)
A collection of requirements for a standardized M2M (Machine to Machine) interface that permits utilities to share, at the consumer's authorization, a broad set of that consumer's utility data with third parties. It is also a task force within the UCAIug (Utility Communications Architecture International Users Group) Open Smart Grid subcommittee that is building interoperable data exchanges.
(http://osgug.ucaiug.org/sgsystems/OpenADE/default.aspx)

OpenADR (Open Automated Demand Response)
A suite of functions and capabilities that facilitates continuous, bi-directional and secure DR communications between utilities or CSPs (Curtailment Service Providers) and their customers. The specification uses open and non-proprietary Internet–based price, reliability, or event signals to automated commercial and industrial energy management and

residential control systems to support retail and wholesale applications. Version 1.0 is released, and Version 2.0 is in development in partnership with the NIST (National Institute of Standards and Technology) Smart Grid Interoperability Standards activities. (http://openadr.lbl.gov/)

OpenADR Alliance
A nonprofit association of government and industry focused on establishing a national ADR standard using OpenADR through education, training, testing, and certification. (http://www.openadr.org/FAQs.aspx)

OpenAMI (Open Advanced Metering Infrastructure)
A global task force of major vendors, utilities, researchers, and consultants from across the electrical metering industry whose goal is to establish an open standard communications architecture for metering and demand response applications. It focuses on applying open standards to two functions: automatic metering and demand response. The task force is part of the UCAIug (Utility Communications Architecture International Users Group). (http://osgug.ucaiug.org/OpenAMI/default.aspx)

OpenESPI (Energy Service Providers Interface)
A project whose mission is to provide an open source implementation of the NAESB REQ.21 Energy Service Provider Interface standard. OpenESPI includes two major players; A data custodian and a third party. These two interact to both test standard implementation and direct ESPI conformance tests. OpenESPI is designed to meet the implementation and conformance testing requirements of the UCAIug OpenADE Task Force. The project is involved in the Green Button initiative. (http://www.openespi.org/)

OpenHAN (Open Home Area Network)
A task force under the guidance of the UtilityAMI Working Group, which is part of the UCAIug (Utility Communications Architecture International Users Group) and the Open SG (Open Smart Grid) subcommittee. It developed the UtilityAMI 2008 Home Area Network System Requirements

Specification or UtilityAMI 2008 HAN SRS. This specification targets standards and technology development for HAN systems interoperability with utilities and consumer portals, such as a meter. (http://osgug.ucaiug.org/sgsystems/openhan/default.aspx)

OpenGIS®
Standards developed by OGC (Open Geospatial Consortium) to enable interoperability between various geoprocessing technologies that enable the Web, wireless and location-based services, and other applications to leverage complex spatial information and services.

OpenRF™ (Open Radio Frequency)
An embedded wireless platform that provides a tree-mesh networking protocol and an event-driven application framework for HAN (Home Area Network) applications.

OpenSG (Open Smart Grid)
An administrative subcommittee of the UCAIug (Utility Communications Architecture International Users Group) that sponsors working groups to address Smart Grid requirements and develop interoperability guidelines. It is one of three subgroups in UCAIug. It oversees the following Working Groups: Security (UtiliSec), Communications (UtiliComm), Systems, and CWG (Conformity Working Group). (http://osgug.ucaiug.org/default.aspx)

Operating reserve
The power required for regulation, load forecasting errors, equipment outages, and local area protection beyond the expected system demand. Operating reserve provides a safety margin that helps ensure reliable electricity supply. It consists of spinning and non-spinning reserve.

Operating revenue
The revenue generated from electricity sales to customers. It is collected through rates that may consist of a number of separate components,

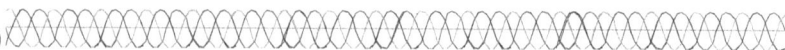

including energy charges, demand charges, consumer service charges, environmental surcharges, and fuel and purchased power adjustments. Costs that vary with the amount of electricity produced are generally recovered through energy charges. Costs that do not vary with production, such as capital costs, are recovered through demand charges.

OPGW (OPtical Ground Wire)
A cable used for electric transmission grids that can shield or ground high voltage lines and deliver communication capabilities in a dual function cable arrangement of metal and optical fiber.

OPS (Office of Pipeline Safety)
A PHMSA (Pipeline and Hazardous Materials Safety) department that has a national program to ensure the safe, reliable and environmentally sound operation of the USA's pipeline transport system. (http://phmsa.dot.gov/portal/site/PHMSA/menuitem.ebdc7a8a7e39f2e55c f2031050248a0c/?vgnextoid=ca9fe4fca0380110VgnVCM100000762c7798R CRD&vgnextchannel=8938143389d8c010VgnVCM1000008049a8c0RCRD&v gnextfmt=print)

Order 693
A FERC (Federal Energy Regulatory Commission) rule issued in 2007 that established 83 mandatory reliability standards for the bulk-power system.

Order 706
A FERC (Federal Energy Regulatory Commission) rule issued in 2008 with eight reliability standards for Critical Infrastructure Protection by NERC (North American Electric Reliability Corporation) to safeguard critical cyber assets.

Order 719
A FERC (Federal Energy Regulatory Commission) rule issued in 2008 that, among other things, directs ISOs and RTOs to accept bids from demand

response resources in their competitively bid markets in ancillary service markets. It also addressed long-term power contracting, market-monitoring policies; and RTO and ISO responsiveness to stakeholders and customers. The objectives are to enable more supply options, improve operating performance, and encourage technology innovations.

Order 888

A FERC (Federal Energy Regulatory Commission) rule issued in 1992 ordering utilities to open their transmission grids to unaffiliated generators. Order 888 requires that the utilities "functionally unbundle" their generation and transmission businesses to prevent anticompetitive favoritism in granting access to affiliated electricity producers. The resulting typical organization has been the ISO (Independent System Operator), whereby utilities continue transmission facility ownership but give operational authority to an independent board. Order 888 includes suggested rules for ISOs, and FERC retains authority to approve or disapprove the specific procedures ISOs choose.

Order 889

A 1996 FERC (Federal Energy Regulatory Commission) order that had a key role in opening the USA energy market to competition. It set standards for information that must be available to the marketplace and established OASIS (Open Access Same-time Information System) to share this information. This order requires utilities to share market information and make it available to potential competitors through OASIS.

Order 890

This FERC (Federal Energy Regulatory Commission) order identified reforms to the *pro forma* OATT (Open Access Transmission Tariff) in the areas of calculation of available transfer capability, transmission planning, and the terms and conditions of open access transmission service. It increased transparency in the interconnection transmission planning processes.

Order 1000
A FERC (Federal Energy Regulatory Commission) rule that builds on Order 890 and creates more regionally-based transmission planning processes and regionally-based cost allocation methods. It also eliminates the federal right of first refusal in certain circumstances. Public utility transmission providers must adhere to these processes.

Order 2000
A FERC (Federal Energy Regulatory Commission) rule that encouraged utilities, which own interstate transmissions facilities, to participate in an RTO (Regional Transmission Organization), to promote efficiency in wholesale electricity markets, and to ensure the lowest prices for reliable service. This rule also outlined 12 characteristics and functions that an entity must satisfy to become an RTO.

ORMM (Outage Restoration Maturity Model)
A model built by industry consensus that identifies crucial outage management and restoration functions. Using this model as a tool, utilities can benchmark internal operations for outage restoration and compare to industry best practices.

ORNL (Oak Ridge National Laboratory)
The DOE's (Department of Energy) largest science and energy laboratory. Managed since April 2000 by a partnership of the University of Tennessee and Battelle, it has six major mission roles that include energy, high-performance computing, and materials science at the nanoscale. Energy initiatives include hosting the USA project office for the ITER international fusion experiment and the Energy Efficiency and Renewable Energy (EERE) Program, which focuses research in transportation, industrial technologies to cut energy use, buildings and weatherization, and various renewable energy sources. One program is the Sustainable Electricity Program for EV battery reuse. (http://www.ornl.gov/)

OSGi$^{(TM)}$ Alliance (Open Services Gateway Initiative Alliance)
A global industry consortium that promotes JAVA-based specifications, reference implementations, test suites, and certification to assure interoperability of applications and services based on its component integration platform. (www.osgi.org)

OSHPD (Office of Statewide Health Planning and Development)
A California state agency that requires certification that power generation equipment will survive seismic events and operate adequately to meet building performance needs. (http://www.oshpd.ca.gov)

OSI (Open Systems Interconnection)
Also known as the OSI Model, a reference model that describes how any two devices communicate in a network. It consists of seven layers and has been adopted by the ITU (International Telecommunications Union) as Recommendation X.200. The layers are Physical, Data-link, Network, Transport, Session, Presentation, and Application.

OSIAN (Open Source IPv6 Automation Network)
A free and downloadable wireless network that operates in the 900-MHz spectrum using IEEE (Institute of Electrical and Electronics Engineers) 802.15.4 with 6LoWPAN (IPv6 Low power Wireless Personal Area Network) for IP addressing and support for the ZigBee protocol. It is characterized as a low-power, secure, and robust network capable of penetrating walls and other barriers that challenge higher-frequency networks.

OTA (Over The Air antenna performance)
The measurable efficiency of transmission and reception of an antenna from active wireless devices such as mobile phones or smart meters. Criteria for excellent or poor OTA are based upon transmitting power and receiving sensitivity in an environment with normal impairments such as close proximity physical objects.

OTA Update
This is the wireless delivery methodology for software updates and data for mobile phones, tablets, and other smart devices. The most common use for OTA updating is to deploy new operating systems for remote and mobile devices.

OTC (Once-Through Cooling)
Used in context with electricity generation, water that is withdrawn from a coastal or estuarine source, circulated through the power plant infrastructure, and then returned to a water body at a higher temperature. Cooling water withdrawals cause adverse impacts when larger aquatic organisms, such as fish and mammals, are trapped against a facility's intake screens and when smaller organisms, such as larvae and eggs, are drawn through the cooling system and killed. In California, the State Water Resources Control Board established technology-based standards to implement federal Clean Water Act section 316(b) and reduce the harmful effects associated with cooling-water intake structures on marine and estuarine life. The policy applies to nineteen existing power plants (including two nuclear plants) that currently have the ability to withdraw over 15 billion gallons per day from the state's coastal and estuarine waters using a single-pass system, also known as once-through cooling.

Outage
The time period when some part of the T&D (Transmission and Distribution) system is not functioning for either planned reasons, like repairs, or unplanned reasons, like weather or component failure.

Outage detection
The capability to automatically communicate outages from meters or substations back to a utility to expedite fault isolation and repair work.

Outfall
The location where treated or untreated water from sewer or drain is discharged into the environment.

Overland flow
A wastewater treatment process that distributes wastewater over a land slope for natural filtration and collects the bottom runoff for reuse.

Overvoltage
A temporary increase in voltage that lasts longer than a brief period of time like a second. When it is less than this timeframe, it is typically a voltage swell.

OW (Office of Water)
Part of the EPA responsible for implementing the federal government's oversight of the Clean Water Act, Safe Drinking Water Act; and other water-related legislation. It provides guidance, stipulates scientific methods and data collection requirements and conducts oversight. Each of EPA's ten regions has a Water Division that implements all programs. The OW also contains the following Offices: American Indian Environmental Office; Office of Wastewater Management; Office of Science and Technology; Office of Ground Water and Drinking Water; and the Office of Wetlands, Oceans and Watersheds.
(http://water.epa.gov/aboutow/org/programs/owintro.cfm)

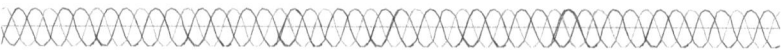

P

P-Class type synchrophasor
Devices dedicated to capturing protection data instead of measurement data. Protection data is extremely time critical.

P25 (Project 25)
An evolving set of standards for digital wireless bi-directional products used in public safety. It is based in part on TIA-102 standards suite.

P2D (Prices to Devices)
An abbreviated description of the technology and process in which dynamic utility price signals are sent in advance or in realtime to customer equipment and appliances, causing automatic adjustment of their current or scheduled operation to reduce peak demand. Utility P2D communications could connect to building energy management systems and smart devices.

Pacific Northwest Demand Response Project
A project formed in 2007 to encourage the development of DR (Demand Response) programs in the Pacific Northwest. The project is supported by four states in the Pacific Northwest, Bonneville Power Administration, utilities, and the Northwest Power and Conservation Council. (http://www.nwcouncil.org/energy/dr/Default.asp)

PAFC (Phosphoric-Acid Fuel Cell)
Fuel cell technology with applications in stationary power generation systems, including CHP (Combined Heat and Power) systems. It is a mature fuel cell technology, consisting of a liquid phosphoric acid electrolyte, and requires hydrogen as a fuel.

PAN (Personal Area Network)
An individual personal network consisting of devices, such as computers, phones, and peripherals, connected by some type of wireless technology. These networks usually have a range of about 30 feet or 10 meters.

PANA (Protocol for Carrying Authentication for Network Access)
An IP-based standard published by the IETF to provide a uniform authentication protocol across different access technologies such as WiFi, WiMAX, and Ethernet.

PAP (Priority Action Plan)
A set of priority actions based on the recommendations of NIST (National Institute of Standards and Technology) Smart Grid standards efforts. These priority actions result in plans that define problems, establish objectives, and identify the likely standards bodies and associations required for standards modifications, enhancements, and harmonization. There are currently 18 PAPs in various stages of completion

PAP 0 (Priority Action Plan 0) Meter Upgradeability Standard
This standard was completed to define requirements for secure local and remote firmware upgrades of smart meters.
(http://collaborate.nist.gov/twiki-sggrid/bin/view/SmartGrid/PAP00MeterUpgradability)

PAP 1 (Priority Action Plan 1) Role of IP in the Smart Grid
An investigation into the role of IP (Internet Protocol) network technologies in the Smart Grid that published a core Internet Protocol Suite for IP-based Smart Grid in 2010. (http://collaborate.nist.gov/twiki-sggrid/bin/view/SmartGrid/PAP01InternetProfile)

PAP 2 (Priority Action Plan 2) Wireless Communications for the Smart Grid
An ongoing assessment of the capabilities of licensed and unlicensed spectrum wireless technologies to develop guidelines on their use for

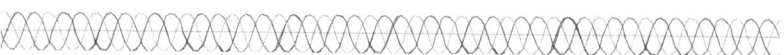

different Smart Grid application requirements.
(http://collaborate.nist.gov/twiki-
sggrid/bin/view/SmartGrid/PAP02Wireless)

PAP 3 (Priority Action Plan 3) Develop Common Price Communication
Model
Ongoing work to create a common price model that works for wholesale
and retail electricity markets and can accommodate distributed energy
resources, "prices to devices", and Demand Response program
communication flow. It overlaps with other PAPs that include price and
product information (4, 6, 8, 9, 10, and 11).
(http://collaborate.nist.gov/twiki-
sggrid/bin/view/SmartGrid/PAP03PriceProduct)

PAP 4 (Priority Action Plan 4) Develop Common Schedule Communication
Mechanism for Energy Transactions
A standard defining how schedule and event information is passed between
and within services that can then be incorporated into price, demand-
response, and other specifications. (http://collaborate.nist.gov/twiki-
sggrid/bin/view/SmartGrid/PAP04Schedules)

PAP 5 (Priority Action Plan 5) Standard Meter Data Profiles
The objective is to develop a smaller set of data tables than currently
supplied in ANSI C12.19 to simplify the meter procurement process and aid
vendors developing products that may require access to local meter data.
(http://collaborate.nist.gov/twiki-
sggrid/bin/view/SmartGrid/PAP05MeterProfiles)

PAP 6 (Priority Action Plan 6) Common Semantic Model for Meter Data
Tables
This PAP is producing a comparison of ANSI C12.19-2008 and IEC 61968-9
meter data models to make it easier for utilities to manage meters from

multiple vendors and ensure common data is available to MDMS and other applications without implementation of translation gateways. (http://collaborate.nist.gov/twiki-sggrid/bin/view/SmartGrid/PAP06Meter)

PAP 7 (Priority Action Plan 7) Electric Storage Interconnection Guidelines
This PAP focuses on electrical interconnection standards, communication standards, and implementation guidelines for energy storage devices that include DER (Distributed Energy Resources) and EV (Electric Vehicles) for energy storage. (http://collaborate.nist.gov/twiki-sggrid/bin/view/SmartGrid/PAP07Storage)

PAP 8 (Priority Action Plan 8) Common Information Model/61850 for Distribution Grid Management
This PAP focuses on the information support of DA (Distribution Automation), back-office applications, and application-based information exchanges relying on IEC 61968 CIM development focus. DA interactions involve monitored and controlled devices serving as actuators and information sources such as voltage and VAR controllers, remotely controlled switching devices, and distributed energy resources. (http://collaborate.nist.gov/twiki-sggrid/bin/view/SmartGrid/PAP08DistrObjMultispeak)

PAP 9 (Priority Action Plan 9) Standard DR (Demand Response) and DER (Distributed Energy Resources) Signals
This PAP defines consistent signal semantics for DR and DER devices. Consistent signals accelerate automation of the DR value chain and increase the ability to use DER devices for balancing reserve, spinning reserve, and other techniques for successful integration to the grid. (http://collaborate.nist.gov/twiki-sggrid/bin/view/SmartGrid/PAP09DRDER)

PAP 10 (Priority Action Plan 10) Standard Energy Usage Information
This PAP completed data standards to exchange fine grained and timely information about energy usage at the consumer level – residential, commercial, and industrial.
(http://collaborate.nist.gov/twiki-sggrid/bin/view/SmartGrid/PAP10EnergyUsagetoEMS)

PAP 11 (Priority Action Plan 11) Common Object Models for Electric Transportation
This PAP developed interoperability standards for PEVs (Plug-in Electric Vehicles) that optimize charging capabilities and encourage adoption of these vehicles. It also supports energy storage integration with the distribution grid addressed by PAP 7. (http://collaborate.nist.gov/twiki-sggrid/bin/view/SmartGrid/PAP11PEV)

PAP 12 (Priority Action Plan 12) Mapping IEEE 1815 (DNP3) to IEC 61850 Objects
This PAP is focused on power delivery communication standards – specifically developing IEEE 1815 (DNP3) and IEC 61850 mapping documents with guidelines to achieve interoperable integration of equipment using DNP3 with equipment using IEC 61850.
(http://collaborate.nist.gov/twiki-sggrid/bin/view/SmartGrid/PAP12DNP361850)

PAP 13 (Priority Action Plan 13) Harmonization of IEEE C37.118 with IEC 61850 and Precision Time Synchronization
IEEE Standard C37.118 is the current primary standard for the communications of PMUs (Phasor Measurement Units) and PDC (Phasor Data Concentrator) data, but the use of IEC 61850 for wide-area communication can impact phasor measurement and applications. This PAP focuses on integration of PMU and PDC data based on C37.118 into IEC 61850.

(http://collaborate.nist.gov/twiki-
sggrid/bin/view/SmartGrid/PAP1361850C27118HarmSynch)

PAP 14 (Priority Action Plan 14) Transmission and Distribution Power
Systems Model Mapping
This PAP defines strategies for integrating several standards across
different environments to support different real-time and back-office
applications. The standards include IEC 61850, IEC 61968 and IEC 61970.
(http://collaborate.nist.gov/twiki-
sggrid/bin/view/SmartGrid/PAP14TDModels)

PAP 15 (Priority Action Plan 15) Harmonize Power Line Carrier Standards
for Appliance Communications in the Home
This PAP is focused on the development of an interoperable profile
containing common features for home appliance applications that use
powerline communications.
(http://collaborate.nist.gov/twiki-
sggrid/bin/view/SmartGrid/PAP15PLCForLowBitRates)

PAP 16 (Priority Action Plan 16) Wind Plant Communications
This PAP is focused on adoption of an international wind power
communications standard in the USA as is done in other parts of the world.
(http://collaborate.nist.gov/twiki-
sggrid/bin/view/SmartGrid/PAP16WindPlantCommunications)

PAP 17 (Priority Action Plan 17) Facility Smart Grid Information Standard
This PAP activity will lead to a data standard that provides a basis for
common information exchange between control systems and end use
devices found today primarily in commercial and industrial facilities. The
common information exchange will describe, manage, and communicate
information on aggregate electrical energy consumption and forecasts.
(http://collaborate.nist.gov/twiki-
sggrid/bin/view/SmartGrid/PAP17FacilitySmartGridInformationStandard)

Parasitic loss
Energy that is consumed by an appliance (ex: computer, WiFi router, tea kettle) that does not go towards the final service provided by the appliance. "Stand-by" mode is an example of this, where electricity is consumed even when the appliance is not actively operating. In the tea kettle example, the energy required simply to keep the kettle at an elevated temperature instead of heating the water further towards boiling is also parasitic loss.

Passive House
A design concept with the objective of primarily heating buildings by passive solar gain and cooling by shading and window orientation. It is different from zero net energy in its reliance on materials and design to reduce energy needs.

PCS (Power Conditioning System)
Used in connection with energy storage and distributed generation sources, technologies that aid in the transfer of energy from these sources back to the grid. PCS acts as a "filter" to eliminate reactive power demand and distortions in harmonic currents. PCS can also refer to products that eliminate power disruptions, surges and spikes coming from the grid to buildings or devices.

PCT (Participant Cost Test)
A type of test used by utilities and regulatory agencies to evaluate the cost-effectiveness of energy efficiency programs and portfolios. It measures the benefits to rate payers who participate in energy efficiency programs.

PCT (Programmable Communicating Thermostat)
A thermostat that controls HVAC components based on user time and temperature preferences and can communicate with the utility. It can be used in demand response programs to modify demand during peak times.

PDC (Phasor Data Concentrator)
A logical unit that collects data from PMUs (Phasor Measurement Units) and transmits it to other applications. It may temporarily buffer information but does not store this data.

PEAC (Power Electronic Application Center)
An EPRI (Electric Power Research Institute) laboratory in Knoxville, Tennessee with a research focus that includes power quality problems and interconnection of distributed generation or energy resources on the grid. It has a facility that evaluates distributed energy resource compatibility under varying conditions called the Power Quality Distributed Resources or PQDR Park. It is also involved in the development of certifications for grid compatibility. (http://www.epri-peac.com/)

Peak demand
The maximum electricity load required for a period of time, which can be a specific point in time or averaged over a period of time. Also known as Peak Load.

Peak flow
The highest volume of water that a wastewater treatment expects to receive at any point in time.

Peaker plant
A generator that is reserved for use during highest peak load periods. Also known as a Peak Load Plant.

Peaking capacity
The total capacity of generators within a utility's T&D (Transmission and Distribution) system.

PEEC (Precourt Energy Efficiency Center)
A Stanford University research institute promoting energy-efficient technologies, systems, and practices,. PEEC works to understand and overcome market, policy, technology, and human behavioral barriers to reductions of energy use and inform public and private policymaking. (http://peec.stanford.edu/index.php)

PEMFC (Polymer Electrolyte Membrane Fuel Cell)
An automotive fuel cell, also called a Proton Exchange Membrane fuel cell that uses hydrogen fuel and oxygen to produce electricity to power a vehicle.

Penetration testing
A process to identify vulnerabilities in computer networks or applications to improve security measures. Also known as pen testing or ethical hacking.

Performance-based regulation
A rate-setting tactic used by state regulatory agencies. It creates financial motivation for utilities to perform to metrics that could include DSM (Demand-Side Management) activities, customer satisfaction, cost reductions, and power outage factors. It is an alternative to cost-of-service regulation.

Personal Information
Data about an individual that identifies her or him, including name, birthdate, address, credit card history, and other unique information that describes that individual.

PET (Privacy Enhancing Technologies)
Technologies that protect or enhance an individual's physical or data privacy, help enable individual or organizational compliance with privacy policies, or help protect against breaches of privacy. Some examples of PETs include encryption and spam filters.

PEV (Plugin Electric Vehicle)
Any vehicle that contains a plug for external charging of a battery that stores electricity for propulsion. PEVs may also contain an on-board electrical generator.

PEV Collaborative
A public-private effort to enable transitions to a plug-in electric vehicle market in California. It endorses a systematic, customer-focused approach to foster customer adoption of EVs (Electric Vehicle). Members include elected and appointed officials, automakers, utilities, infrastructure providers, environmental organizations and others. (http://www.evcollaborative.org/)

PHA (Passive House Alliance)
An organization that provides education and advocates for policy and code issues, and works with building materials suppliers to meet Passive House standards. (http://www.phalliance.com/home-page)

Pharming
A cyber security threat whereby malicious code is installed on a computer that directs users to unauthorized Web sites.

Phase voltage
The voltage between a conductor and neutral, or ground. Phase voltage can be found by dividing the Line Voltage by the square root of three.

Phasor
A complex number that includes the phase angle and magnitude of voltage and frequency at a specific point in time. It is the measurement gathered by a PMU (Phasor Measurement Unit).

PHEV (Plugin Hybrid Electric Vehicle)
A vehicle that combines characteristics of hybrid and EV vehicles, but has larger batteries and can recharge from a standard outlet. It uses electricity for short trips and daily commutes, but an internal combustion engine for longer distances. The battery in a hybrid is charged internally by electricity generated by the engine and electricity from regenerative braking. From a utility's perspective, a PHEV is equivalent to one-third of a home, approximately 330 KW. PHEVs may play a role in voltage regulation to help reduce the intermittency of wind power.

PHEV roaming charge (Plugin Hybrid Electric Vehicle roaming charge)
A concept similar to cell phone roaming charges that delivers a defined billing structure to EV charging regardless of location, so that the vehicle owner, rather than the plug owner, pays the charge.

PHIUS (Passive House Institute US)
A nonprofit organization that provides training, education, and research to promote implementation of Passive House building energy standard and the design approach and techniques to achieve it. (http://www.passivehouse.us/passiveHouse/PHIUSHome.html)

PHM (Prognostic Health Management)
Technologies that observe systems or components, sense abnormalities, and make predictions of future failures. PHM reduces maintenance costs and improves the available time for monitored devices or systems.

PHMSA (Pipeline and Hazardous Materials Safety Administration)
An agency within the U.S. Department of Transportation with oversight of the transport of hazardous materials by any transport modes, including pipelines. (http://www.phmsa.dot.gov/)

Physical interface
As defined in the ARRA Section 1306 amending EISA 2007 (Energy Independence and Security Act of 2007), a standard physical configuration specification that facilitates the communication functions of a logical interface.

PIA (Privacy Impact Assessment)
A process used to evaluate information privacy when planning, developing, implementing, and operating enterprise information management systems that maintain information on individuals. It consists of privacy training, gathering data from a project on privacy issues, and identifying and resolving the privacy risks.

Picogrid
A microgrid scaled to a neighborhood that includes its own generation, transmission, distribution, energy storage, and energy management software that is connected to a utility power system, but can operate independently from the grid for defined periods of time to respond to load shedding or unplanned outages.

PIER (Public Interest Energy Research)
A CEC (California Energy Commission) program that provides recommendations and funding to advance RD&D projects in energy efficiency and demand response, renewable resources, generation, transmission, distribution, and transportation technologies. (http://www.energy.ca.gov/research/annual_reports.html)

PIG (Pipeline Insertable Gauge)
A mobile tool with one or more measurement sensors or devices placed in a pipeline to record and/or transmit measurements as it moves in that pipe. Specialized utility pigs perform functions such as cleaning pipelines. Others include cameras, ultrasonic measurement equipment, and configuration measures.

PII (Personally Identifiable Information)
Data about an individual that identifies him or her, including name, birthdate, address, credit card history, and other unique information that describes that individual. Also known as Personal Information.

Pipeline integrity management
A formal program to establish, monitor, and manage safe transmission of natural gas. It is comparable to process safety management programs used in industrial facilities that assess risks and develop mitigation plans to reduce those risks.

Pipeline Safety Improvement Act of 2002
Federal legislation that enacted a one call for excavation program and created provisions regarding risk analysis and integrity management programs for gas pipelines. It is under the auspices of the Department of Transportation.

PIPES (Pipeline Inspection, Protection, Enforcement, and Safety Act)
A 2006 federal law that established that operators of natural gas distribution lines must implement enhanced safety programs similar to those already in place for natural gas transmission lines. Key elements include development and implementation of a written integrity management plan; identification of existing and potential threats; assessment, prioritization and mitigation of risks; and performance measurement, monitoring and reporting to regulators. It also amended the 2002 Pipeline Safety Improvement Act to include R&D and standardization activities related to pipeline defect detection and best practices to manage safety risks.

PJM (Pennsylvania, New Jersey, and Maryland Interconnection)
PJM is an RTO (Regional Transmission Organization) that manages the high-voltage grid and wholesale electricity market in all or parts of 13 states and the District of Columbia. It covers all or most of Pennsylvania, New Jersey,

Maryland, Delaware, the District of Columbia, Virginia, West Virginia, and Ohio, and parts of Illinois, Michigan, Indiana, Kentucky, North Carolina, and Tennessee, or 168,500 square miles, with a population of about 51 million. It is one of the 10 ISO (Independent System Operators) or RTOs currently operating in North America. (http://www.pjm.com/)

Planning coordinator
One of the regional functions that contributes to the reliable planning and operation of the bulk power system. Planning coordinators are responsible for regional planning, including assessing the longer-term reliability of the bulk power system, in coordination with other regions.

PLC (Peak Load Capacity)
A demand response program that defines the amount of reduced electricity consumption for a current year based on prior year energy use.

PLC (Power Line Carrier)
A communication system in which the utility power line is the communication link to transmit meter data and other utility system data. Some smart meter and HAN (Home Area Network) solutions use the power line to carry bi-directional communications. It is characterized as very-low bandwidth and capable of longer-distance transmissions than BPL (Broadband Powerline).

PLC (Programmable Logic Controller)
A type of specialized computer for industrial applications that accept data from sensors and switches and perform commands – usually on/off actions. Computer programming logic within a PLC contains the commands and conditions for activation.

PLMA (Peak Load Management Alliance)
A not-for-profit organization consisting of electricity suppliers, load-shedding systems manufacturers, consultants, research groups, and trade

associations to promote the concepts and technologies of demand response using pricing signals. (http://www.peaklma.com/)

Plug In America
A nonprofit organization of electric vehicle owners and advocates for clean air and energy independence and promotes the development and use of battery and plug-in hybrid vehicles. (http://www.pluginamerica.org/)

Plug valve
A valve generally used in wastewater plants that has an extremely tight seal and rugged design.

PM (Phase modulation)
A technique for transmitting information by radio carrier waves in which the phase of the current waveform is changed.

PMA (Power Marketing Administration)
Federal entities within the DOE (Department of Energy) that market power produced at one of four federal hydro projects: Bonneville Power Administration, Southeastern Power Administration, Southwestern Power Administration, and Western Area Power Administration.

PMA (Power Matters Alliance)
An international not-for-profit industry association that promotes interoperable wireless charging technologies in a diverse set of industries, most notably telecommunication and portable consumer electronic devices. (http://www.powermatters.org)

PMU (Phasor Measurement Unit)
A real-time measurement of positive-sequence voltages and currents to deliver dynamic-grid situational awareness. Grid condition sampling occurs at 30 samples/second, with measurements time-stamped to a common GPS time synchronization signal. (In contrast, SCADA (Supervisory Control

And Data Acquisition) systems measure every 4 seconds.) PMU voltage magnitude and phase angle measurements must conform to IEEE (Institute of Electrical and Electronics Engineers) C37.118-2005 requirements. Collection of data from substations fitted with PMUs provides a good snapshot of the state of the power system. PMUs take measurements called synchrophasors. PMUs can be used for automating power systems, increasing grid reliability, and assisting in DR (Demand Response) management. An example of a PMU is a digital relay.

PMU Registry
A NERC (North America Electric Reliability Corporation) initiative that documents the configuration of phasor devices on the North American electrical grid. It provides a centralized structure for mapping PMU data across all utilities and standard naming convention for devices. (https://naspi.tva.com/pmuregistry/)

PNNL (Pacific Northwest National Lab)
A DOE (Department of Energy) laboratory operated by Battelle with a number of research missions. These include improving grid reliability and productivity with tools for wide area monitoring and real-time analysis of grid operations. It is participating in a GridWise™ demonstration to test human response to price signals. The lab developed a circuit called a GridFriendly™ controller that can be used with home appliances. It also has a functional control center for use as an R&D platform, called the Electricity Infrastructure Operations Center or EIOC that provides access to real data from eastern and western power grids. It is working with LBNL (Lawrence Berkeley National Laboratory) and SNLA (Sandia National Laboratory) on mathematical analysis related to the Smart Grid. Other R&D is concentrated in automated buildings diagnostics and control, energy codes and standards, market transformation, and energy program design and implementation. (http://www.pnl.gov/)

PODS (Pipeline Open Data Standards[TM])
A database architecture specific to pipeline operations to store and analyze geospatial data, asset information, regulatory compliance data, and operations data for management of natural gas and hazardous liquids pipelines.

PODS Association (Pipeline Open Data Standards[TM] Association)
A not-for-profit industry association that develops and maintains the PODS Data Model, along with pipeline data storage and interchange standards to address these data management needs of pipeline operators. (http://www.pods.org/)

Port electrification
A program in early implementation stages to reduce CO2 emissions at shipping ports through a number of tactics. Top on the list is providing docked ship plug-ins to the port or shore plugs to eliminate 24X7 idling of ship diesel engines. Second is electrification of port vehicles.

Portfolio Manager Overview
A tool developed by the US EPA that is based on CBECS to evaluate building energy consumption performance. (www.energystar.gov/index.dfm?c=evaluate_performance.bus_portfoliomanager)

Potable water
Water that is safe for human consumption.

Potential peak reduction
The DR (Demand Reduction) that is possible during events that trigger requests to curtail load. It is not the actual peak reduction. For utilities, the calculation is the potential sum of the DR capability to annual MW peak load achieved by the program participants. For ISOs (Independent System Operators) and RTOs (Regional Transmission Organizations), the calculation

is the sum of coincident reduction capability achieved by participants at the time of ISO/RTO system peak. For CSPs (Curtailment Service Providers), the calculation is the sum of coincident reduction capability that is met by their DR program participants at the peak time in the region.

POTWs (Publicly-Owned Treatment Works)
Wastewater treatment units, sewers, sewage collection systems, pumping, power, and other equipment or facilities that are publicly owned.

Power 2.0
A suite of standards under development that address the consumer power ecosystem, including mobile, computing, car-based, appliances and electronics, smart batteries, household power, and power-in-public-places. Power 2.0 will also create specifications for the interaction of software and services with power. The Power Matters Association is defining these standards.

Power conversion and efficiency
A ratio applied to devices that describes output power to input power. For example, an inverter loses energy converting DC to AC, so this ratio helps identify the most efficient device. For a number of devices, the lost energy is converted into heat. Also known as power consumption.

Power electronics
A field of technical knowledge that focuses on processing and controlling electric power over long distances using electronic devices. Equipment examples include converters, inverters, and rectifiers in transmission systems.

Power exchange
An entity providing a competitive spot market for electric power through day and/or hour-ahead auction of generation and demand bids.

Power marketer
A business entity engaged in buying and selling electricity. Power marketers typically do not own generating or transmission facilities. Power marketers, as opposed to brokers, take ownership of the electricity and are involved in interstate trade. These entities must file with FERC (Federal Energy Regulatory Commission) to obtain status as a power marketer.

Power pool
Two or more interconnected utilities that pool their resources to reliably and economically supply the power and energy requirements for both utilities' customers. Resources may include generating facilities, transmission system access, and back-office administrative applications.

Power quality
An assessment of the stability of a grid or assets to deliver electricity without voltage sags, surges, or other disruptions in voltage, frequency, or power factors.

Power quality monitoring
The ability of the AMI (Advanced Metering Infrastructure) network to discern, record, and send back to the utility any instances in which the voltage and/or frequency were not in acceptable ranges for reliability.

Power systems management
A management system that includes operations within control centers, substations, and equipment, including telecontrol and monitoring interfaces to equipment, systems, and databases.

Power tower
An array of mirrors called heliostats that track the sun and concentrate solar radiation to a central heat exchanger or receiver on a tower to produce steam for electricity. It is used for utility-scale electricity generation from 30 to 400 MW.

PowerMatcher
A distributed energy systems architecture and software. It enables integration of high numbers of distributed energy resources such as small scale generators and PV assets, as well as energy storage and demand response.

PPA (Power Purchase Agreement)
A contract entered into by a power producer and its customers. PPAs require the power producer to take on the risk of supplying power at a specified price for the life of the agreement regardless of price fluctuations.

PPD-8 (Presidential Policy Directive 8)
A presidential directive that focuses on improvements to national preparedness through attention to systemic security and resiliency to security threats.

PPM (Pre-Paid Metering)
Purchase program that puts electricity or gas payments on a smart card. The smart card is inserted in a display unit in the buyer's home. The display unit measures consumption and debits the smart card. Smart cards are "reloaded" with value at designated retail locations.

Ppm (parts per million)
Unit description used with water, it refers to one part per one million parts.

PQR (Power Quality and Reliability)
Two power concepts that are separately measured but closely related in terms of grid performance. Power quality is concerned with maintaining voltage and frequency in acceptable ranges to ensure system reliability. Flicker is one example of poor power quality. Reliability is concerned with total electric interruptions, such as complete loss of voltage and three common metrics: SAIFI (System Average Interruption Frequency Index),

SAIDI (System Average Interruption Duration Index), and CAIDI (Customer Average Interruption Duration Index).

Pressure
The strength of water flow through distribution pipes, measured in pounds per square inch (psi). It is usually supplied by compressed air or gravity.

Pressure vessels
Water storage containers designed to withstand specific psi.

PRIME (PoweRline Intelligent Metering Evolution)
A PLC (Power Line Communications) telecommunications architecture for low voltage lines that works with the CENELEC Band A frequency range. It defines the lower OSI layers.

PRIME Alliance ((PoweRline Intelligent Metering Evolution Alliance)
An industry organization promoting an open and non-proprietary narrowband PLC (Power Line communications) specification and standard, and interoperable equipment for the Smart Grid. (http://www.prime-alliance.org/)

Prime mover
Equipment that converts energy into electricity, such as a steam turbine, internal-combustion engine, solar system, water turbine, or wind turbine.

Privacy
From a Smart Grid perspective, an individual's right to control the collection, use, and sharing of energy (or water) consumption data and other personally identifiable information.

Privacy by Design
A set of principles developed and instituted by the Commissioner of Privacy for Ontario, Canada. These principles are now being reviewed by USA

utilities working on the Smart Grid for potential application in the USA (http://www.privacybydesign.ca/)

Privacy policy
The formal, written statement from an organization regarding collection, use, storage, and destruction of data gathered from consumers or businesses.

Project Haystack
An open source initiative to develop tagging standards and classifications for building equipment and operational data. The product of this initiative is a standardized data model and API for HVAC, lighting, and other environmental systems to communicate over HTTP. (www.project-haystack.org)

Prosumer
A term coined by Alvin Toffler to describe a producing consumer. From a Smart Grid perspective, it would apply to distributed energy resource situations in which the owner of electricity production, storage assets, or participant in a demand response or energy efficiency program may also have a consumer relationship with a utility, aggregator, or other energy services provider.

Protection system
A proposed NERC (North America Electric Reliability Corporation) definition for equipment used to protect generation and transmission assets generally operating at 100kV or higher. It includes protective relays responsive to electrical quantities; communication systems for operation of protective functions; voltage and current sensing devices providing inputs to protective relays; battery backup for protective functions (including station batteries, battery chargers, and non-battery-based dc supply); and control circuitry.

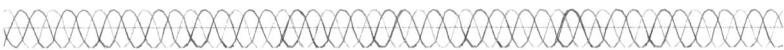

Protocol
A standard set of definitions or rules used in telecommunications to enable data interchange at the hardware level and protocols for data interchange at the application program level. The OSI (Open Systems Interconnection) model identifies one or more protocols at each layer in the telecommunication exchange that both ends of the exchange must recognize and observe.

PSERC (Power Systems Engineering Research Center)
A university/industry partnership focused on R&D for solutions related to power and energy and building educational foundations for power industry careers. There are three primary research areas: systems research, markets research, and T&D (Transmission and Distribution) technologies research. (http://www.pserc.wisc.edu/)

Pseudo generation
Virtual generation resource that has automated curtailment controls to ensure that it has reduced its load in compliance with demand response signals.

PSI (Pounds per Square Inch)
The metric used to describe water pressure.

PSMA (Power Sources Manufacturers Association)
A not-for-profit organization promoting knowledge and education of technological and other developments related to power sources and conversion devices. (http://www.psma.com)

PTCRB (PCS Type Certification Review Board)
A global organization that provides a framework for GSM (Global System for Mobile Communication and UMTS (Universal Mobile Telecommunications System) device certifications. It includes determination of the test specifications and methods necessary to support

the certification process for devices. It authorizes third-party laboratories to conduct testing and is also responsible for input regarding testing of devices to standards development organizations. (http://www.ptcrb.com/)

PTEM (Physical-Technical-Economic Model)
A model for energy-efficiency program design that is focused on technical devices and assumes that all energy users are governed by economic motivations and rational choice.

PTR (Peak Time Rebate)
A payment made to customers who reduce electricity use during peak times. It is considered a carrot-only program since there is no penalty for not reducing demand during peak times.

Public service provider
A public water system owned or managed by a public or cooperative entity such as a city or county.

Public supply withdrawal
Water withdrawn by public or private water suppliers for delivery to domestic, commercial, industrial, and/or hydro power users.

Publish/Subscribe
The current market mechanism that publishes LMP (Locational Marginal Pricing) to interested suppliers and buyers. Smart Grid models may require a more active push mechanism to publish prices to many more users or subscribers.

PUC (Public Utilities Commission)
A common designation for the state regulatory agencies with oversight of IOUs (Investor Owned Utilities), telecommunications, water and sanitation, cable, and other sectors.

PUE (Power Usage Effectiveness)
An energy-efficiency metric that measures how much power goes directly to computing compared to ancillary services, such as lighting and cooling in a data center. It is calculated by dividing the amount of power entering a data center by the power used to run the computer infrastructure within it.

PUHCA (Public Utility Holding Company Act of 1935)
An act written in 1935 that prohibited the acquisition of wholesale or retail electric business by a holding company unless that was part of a public utility entity. It required the parent companies of utilities to be incorporated in the same state that had utility operations for state regulatory purposes or have oversight by the Securities and Exchange Commission (SEC) for utility operations that covered multiple states. It also restricted non-utilities, such as investment banks, to own utilities. This law was repealed in 2005 as part of EPACT 2005.

PUHCA (Public Utility Holding Company Act of 2005)
An act that replaced PUHCA 1935 and revised the definition of holding companies as companies that own 10% or more of the voting securities of a public utility. It provides state and FERC (Federal Energy Regulatory Commission) access to holding company books and records, and defined non-traditional utilities as entities that can include community choice aggregators.

Pumped water storage
Storage facility that receives water pumped up an elevation to generate hydroelectricity at specific times. Also known as pumped hydroelectric storage.

PURPA (Public Utilities Regulatory Policies Act)
An act, written in 1978, that required utilities to connect QFs (Qualifying Facilities) to their transmission grids. The PURPA also required utilities to purchase power from QFs at the avoided cost of additional utility

generators; the states could determine how high the avoided cost was and thus how much outside power utilities would have to buy. The avoided cost was typically the fuel costs incurred in the operation of a traditional power plant, so the Act served the purpose of opening the generation market for renewable sources to compete with carbon-based fuels.

PV (Photo-Voltaic)
A type of solar technology that converts sunlight directly into electricity. PV comprises photo diodes that create DC power, which is changed to AC using inverters.

PWS (Public Water System)
As defined by the EPA, a system that provides water to the public for human consumption through pipes or other constructed conveyances, and has at least 15 service connections or regularly serves at least 25 individuals, for at least 60 days/year. There are approximately 170,000 PWSs in the USA.

Q

QAM (Quadrature Amplitude Modulation)
A method of combining two amplitude-modulated signals to double the effective bandwidth in wireless communications.

QF (Qualifying Facility)
A facility that produces power using renewable fuels, or an industrial cogenerator that produces electricity on site and could supply electricity back to the bulk power system.

QoS (Quality of Service)
A communications term that can be quantified in terms of message delay, message due dates, bit error rates, packet loss, economic cost of transmission, transmission power, and other metrics.

QPSK (Quadrature Phase Shift Keying)
A method of modulation that uses carrier signal phase transitions to increase the amount of information that can be carried in the signal. It is commonly used in wireless communications.

Qualified Smart Grid system
As defined in the ARRA Section 1306 amending EISA 2007 (Energy Independence and Security Act of 2007), any Smart Grid system deployed into service by January 1, 2013 by a utility or energy services provider. It includes, among other things, T&D (Transmission and Distribution) monitoring and communications equipment, meters or sensor and control devices integrated with an electric utility system, retail distributor or marketer, software, and integration controls for distributed generation.

Quick-Start reserve
Reserve capability that can be converted fully into energy within 10 minutes of an ISO (Independent System Operator) request, provided by equipment not electrically synchronized to the power system.

R

Radial feeder
Equipment found in distribution substations that take power from a substation and flow it to customers. Feeder components include transformers, voltage regulators, and capacitors.

Radial grid
A type of electrical distribution grid design. A single wire radiates from the main distribution line to an individual customer. This is typically found in rural areas with isolated loads.

Radius of influence
A term in well hydraulics that describes the radial distance beyond which the aquifer water table is undisturbed by the pumping of water from a well.

RAM-W^{TM} (Risk Assessment Methodology for Water Utilities)
A process created by the EPA in partnership with the Water Research Foundation (WRF) and Sandia National Laboratories (SNL) that is a template and minimum requirements for water utility emergency response plans. It establishes the guidelines for conducting vulnerability assessment programs for all public water utilities serving populations greater than 3,300. (http://cfpub.epa.gov/safewater/watersecurity/glossary.cfm)

Ramp
The up or down change of a generator's output expressed in megawatts per minute, it is often used in discussion of integration of intermittent renewable power sources to the grid.

Ramp event
The dramatic increase or decrease in solar or wind power delivered to an electric grid, based on the variability of these energy sources. The increase

or decrease is significant enough to cause instability to the grid if it is not managed with other sources of power to firm the intermittent sources.

Range anxiety
The fear of many traditional car owners that an EV (Electric Vehicle) cannot support their trip miles, or that a charging station is not within their travel distance. It is considered a primary hurdle to adoption of EVs in the USA

Rankine cycle
Also known as a steam cycle, a thermoelectric generation process in which a fuel is used to heat a liquid (usually water) to produce a high pressure gas that expands over a turbine to produce electricity.

RAP (Regulatory Assistance Project)
A nonprofit organization, formed in 1992 by experienced utility regulators, to provide research, analysis, and educational assistance to public officials on electric utility regulation. Work covers electric utility restructuring, power sector reform, renewable resource development, development of efficient markets, performance-based regulation, demand-side management, and green pricing for domestic and global utility regulators. It also provides regulators with technical assistance, training, and policy research and development. It is an advisor to DOE (Department of Energy) and EPA (Environmental Protection Agency) on the National Action Plan for Energy Efficiency. (http://www.raponline.org/)

RAS (Remedial Action Scheme)
Substation–based processes to protect facilities from overloading under contingency conditions. A RAS consists of monitoring and event detection and mitigation with a local controller within the substation or with neighboring substations. It is also called a SIPS (System Integrity Protection Scheme).

Rate
The price of electricity consumption for a specific time frame. Rates must be approved by a regulatory authority.

Rate base
The value of property on which a utility is allowed to earn a specified rate of return as established by a regulatory authority. Accounting methods to calculate the valuation include fair value, prudent investment, reproduction cost, or original cost.

Rate case
The regulatory process that is used to set utility rates.

Rate features
The description of rate schedules or tariffs offered to utility customers.

Rate-making authority
The organization that is authorized by state or federal legislatures to approve/disapprove rates for a public utility, such as a Public Utility Commission.

Rate of return on rate base
The ratio of net operating income earned by a utility calculated as a percentage of its rate base.

Rate schedule
The description approved by a regulatory agency that covers financial terms and conditions of utility services delivered to customers.

Raw water
Untreated water collected from a surface water source (such as a lake or river) or from a groundwater source (such as a well or aquifer) within the watershed that provides the water resource.

R&D (Research and Development)
The sequence of actions necessary to bring a new solution to a pilot or demonstration stage.

RCD (Remote Connect/Disconnect)
A smart meter capability that enables start and stop of electricity service without requiring physical access to a meter.

RCRA (Resource Conservation and Recovery Act)
Federal legislation passed in 1976 that gives the EPA authority to control all phases of hazardous waste generation, transportation, treatment, storage and disposal, and a framework to address non-hazardous solid wastes such as municipal solid waste from water treatment plants. It was amended in 1986 to include regulation of underground storage of hazardous substances. (http://www.epa.gov/lawsregs/laws/rcra.html)

RCV (Remote Control Valve)
A valve with actuators to open or close the valve based on a signal from a remote location or a control room operator.

RD&D (Research, Development, and Demonstration)
The sequence of actions necessary to bring a new solution to consideration for deployment with a utility.

RDRR (Reliability Demand Response Resource)
A wholesale demand response provider recognized to work in retail emergency-triggered demand response programs within the California ISO market and operations.

RDSI (Renewable and Distributed Systems Integration)
A DOE (Department of Energy) program encouraging public/private partnership demonstration projects of microgrid technologies, distributed generation and energy storage, bi-directional communications, and DR

(Demand Response) programs. The objective is to reduce peak load on distribution feeders by 20 percent by 2015. (www.oe.energy.gov/renewable.htm)

RDT&E (Research, Development, Test, and Evaluation)
A process to create new technologies from research and development that confirms these results via test and evaluation during all stages. It is most often associated with military technology R&D.

RE (Regional Entity)
Eight territorial organizations that supply most of the electricity in North America, plus Baja California in Mexico. These entities or councils are Florida Reliability Coordinating Council, Midwest Reliability Organization, Northeast Power Coordinating Council, Reliability*First* Corporation, SERC Reliability Corporation, Southwest Power Pool Regional Entity, Texas Regional Entity, and Western Electricity Coordinating Council. Regional entities are composed of IOUs (Investor Owned Utilities), federal power agencies, rural electric cooperatives, state, municipal and provincial utilities, independent power producers, power marketers, and end-use customers. Also known as Reliability Councils or Regions.

REA (Rural Electrification Act)
An act passed in 1936 that allowed the federal government to make low-cost loans to not-for-profit cooperatives to deliver electricity to rural areas of the USA. It was part of a stimulus package enacted during the Great Depression, and resulted in widespread electrification of underserved areas.

Reactive power
The portion of electricity that establishes and sustains the electric and magnetic fields of AC equipment. It is provided by generators, synchronous condensers, or capacitors and directly influences electric system voltage. It is usually expressed as kVAR (kilovolt-amperes reactive) or MVAR

(megavolt-ampere reactive), and it is the difference between apparent power and real power.

Reactive power compensation
The deployment of specifically designed devices to address potential grid instabilities and improve AC power system performance. These devices include capacitor banks or static VAR (Volt Amperes Reactive) compensators that reduce reactive power requirements. Use of these devices reduces CO_2 emissions, costs, and transmission losses because it eliminates the need to expand system capacity.

REC (Renewable Energy Certificate)
Issued by government entities to increase production of renewable energy sources. There are two main markets for RECs: voluntary and compliance. Voluntary markets enable customers (residential, commercial, and industrial) to support renewable energy even when it is not available to them as the source of their electricity. Compliance markets exist in the states that require specific percentages of renewable sources of electricity to be generated or sold. Utilities may buy and sell RECs to achieve their percentage if they do not have renewable power generation sources. Also known as Renewable Energy Credits.

Reclaimed wastewater
The reuse of treated wastewater for uses where potable water is not needed – such as irrigation or certain industrial processes.

Recloser
A switch that senses and interrupts faults within the distribution grid and automatically restores service after a momentary outage. There are single-phase and 3-phase reclosers.

Recycled water
Water that is used more than once as potable or nonpotable water before returning back into the natural hydrologic system.

Redox (reduction-oxidation)
A chemical reaction that changes an oxidation state by transferring electrons. It is used in some fuel cell technologies.

REEEP (Renewable Energy and Energy Efficiency Partnership)
Nonprofit focused on promoting renewable energy and energy efficiency in emerging markets and developing countries. (www.REEEP.org)

Regional grid
The existing USA energy grid that comprises three large regional grids or interconnections known as the Eastern grid, Western grid, and Texas grid.

Regulation
The second-by-second adjustment of power production to match voltage loads to instantaneous demand and regulate system frequency in the bulk power system.

Regulation reserve
Generation that is available to support regulation services to maintain the quality of electricity (frequency) in the bulk power system.

Reliability
In the context of the bulk power system, NERC (North American Electric Reliability Corporation) defines reliability as the ability to meet the electricity needs of end-use customers, even when unexpected equipment failures or other factors reduce the amount of available electricity. It is measured by the frequency, duration, and magnitude of events like outages on customers. The two main characteristics of reliability are adequacy and security. From a Smart Grid market perspective, it is dispatchable demand

response and demand-side resources that can supplement online generation.

Reliability coordinator
The highest level of organizational authority responsible for the reliable operation of the bulk power system for realtime operations and next-day analyses, using tools, processes, and procedures with authority to prevent or mitigate emergency operating situations.

Reliability coordinator area
A geographic definition that maps to one or more balancing authority areas. It encompasses the generation, transmission, and loads of the reliability coordinator's region.

Reliable Public Power Provider (RP3)
An achievement program created by the American Public Power Association that encourages public power systems to improve their proficiency in four key areas. The areas are: reliability, safety, workforce development, and system improvement. Once APPA recognizes improvement on these criteria in a public utility, the RP3 designation is awarded.

Relief valve
A valve that gradually opens and closes in response to increased pressure, usually used in systems handling liquids.

Remote fault indicator
Devices that provide information about locations of faults and display or communicate that information to utilities and other asset owners to improve distribution system reliability.

Renewable electricity
As defined in the ARRA Section 1306 amending EISA 2007 (Energy
Independence and Security Act of 2007), electricity derived from a
renewable energy resource, such as wind, solar, biomass, or fuel cells.

Renewable energy resource
As defined in the ARRA Section 1306 amending EISA 2007 (Energy
Independence and Security Act of 2007), energy sources that do not rely on
fossil fuels to generate electricity, such as wind, solar, geothermal, or
biomass.

Renewable resources
Anything that can be reproduced, replaced, or replenished either by natural
cycles or sustainable management. Renewable energy resources include
solar, wind, geothermal, and hydro.

Reporting party
Entities that are responsible for maintaining demand response data in the
NERC (North American Electric Reliability Corporation) system. These
entities, which include balancing authorities, LSEs (Load Serving Entities),
planning authorities, and reliability coordinators, are generally responsible
for dispatching the demand response program, product, or service.

RESA (Retail Energy Supply Association)
An industry association of retail energy suppliers promoting competitive
retail energy markets for residential, commercial, and industrial consumers.
(http://www.resausa.org/)

RESCO (Renewable-based Energy Secure Communities)
A CEC (California Energy Commission) PIER (Public Interest Energy Research)
RD&D initiative using renewable energy technologies with energy efficiency,
Smart Grid integration, energy storage, combined cooling, heating and power,
and co-production of products like biofuels in new communities.

Reservoir
A natural or man-made pond, lake, tank, or basin where potable or nonpotable water is collected and stored.

Resiliency
The fast recovery of an acceptable level of electricity services on a continuing basis despite disruptions to normal operations.

Resource Adequacy
A policy to ensure adequate reserves for reliable electricity service. Several states, notably California, have resource adequacy standards for IOUs (Investor Owned Utilities) and electric service providers to ensure that there is sufficient electricity-generating capacity to meet demand and required reserves during peak demand periods plus a 15-to-18 percent planning reserve margin. This policy assures that there will be adequate resources to provide for state energy supply needs in the year-ahead time frame and beyond.

Resource planner
One of the regional functions contributing to the reliable planning and operation of the bulk power system. Resource planners are responsible for ensuring adequate resources to meet the load in an RTO (Regional Transmission Organization) or ISO (Independent System Operator).

Retail electric supplier
As defined in the ARRA Section 1306 amending EISA 2007 (Energy Independence and Security Act of 2007), an electric utility that has sold at least 1 million MWh of electricity to consumers, as described in PURPA (Public Utilities Regulatory Policies Act) amendments established in 2009.

Retail power
The prices that residential, commercial, and industrial customers actually pay for power. Retail power is typically controlled by regulatory authorities.

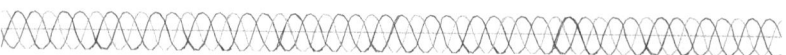

RETI (Renewable Energy Transmission Initiative)
A California statewide initiative to identify transmission projects needed to support renewable energy goals and policy, and facilitate transmission corridor designation and transmission and generation siting and permitting. It will assess competitive renewable energy zones for cost-effective, environmentally sensitive development to supply electricity by 2020. Its members include the CPUC (California Public Utilities Commission), CEC (California Energy Commission), CAISO (California Independent System Operator), and some publicly owned utilities.
(http://www.energy.ca.gov/reti/)

Revenue Requirement
The total amount of money a utility must collect from customers to pay all operating and capital costs, taxes, and debt interest, including a fair return investment.

Reverse osmosis
A water treatment process to filter salty or contaminated water using filters, semi-permeable membranes, and pressure to create potable water. It is used in desalination and wastewater treatment.

RF Mesh Network (Radio Frequency Mesh Network)
A wireless network that is usually designed as a full-access or partial-access topology. Each communicating device is a node or access point, and in a full-access topology, all nodes can talk to each other. In a partial-access topology, points are connected only to some instead of all points. The tradeoffs are reliability and redundancy versus costs and design challenges like LoS (Line of Sight).

RFB (Redox (reduction-oxidation) Flow Battery)
A fuel cell technology that uses liquid electrolytes pumped through reactors to produce energy.

RFC (ReliabilityFirst Corporation)
A not-for-profit company focused on ensuring reliability standards and monitoring standards compliance in the bulk electric system, providing assessments of system reliability, and enhancing security for the interconnected electric systems within the Region. It is one of the eight NERC (North American Electric Reliability Corporation) Regional Entities or reliability regions. ReliabilityFirst companies operate in Indiana, Michigan, Ohio, Pennsylvania, New Jersey, Delaware, Maryland, West Virginia, and parts of Virginia, Illinois, Wisconsin, Kentucky, Tennessee, and the District of Columbia. It is a combination of the ECAR, MAAC, and MAIN (the previous reliability organizations in the region). (http://www.rfirst.org/Default.aspx)

RFC (Request for Comment)
An IETF (Internet Engineering Task Force) document used to develop, discuss, and finalize standards. RFC specifications are accepted as standards after at least one vendor has implemented them.

RFC 6272
This is a request for comment standard produced by the IETF (Internet Engineering Task Force) that identifies the key infrastructure of the IP (Internet Protocol) Suite for use in the Smart Grid. Its objective is to aid users in the construction of IP profiles for the Smart Grid.

RFID (Radio Frequency IDentification)
A wireless technology that uses RFID tags to identify and track assets through radio waves. There are active RFID tags with a battery to transmit information and passive RFID tags that cannot transmit without intervention from an intermediary device (a reader). RFID technology can deliver telemetry information in a Smart Grid.

RFP (Request for Proposal)
A formal process that documents the requirements to be met by a product, service, or solution and requests detailed vendor feedback and pricing information.

RGGI (Regional Greenhouse Gas Initiative)
An initiative launched in 2009 in 10 Northeast and Mid-Atlantic states that is the first mandatory, auction-based effort in the USA to reduce greenhouse gas emissions. The goal is to reduce CO_2 emissions from the power sector 10 percent by 2018. It addresses all power generators with the capacity to produce over 25 MW of electricity. The participating states are Connecticut, Delaware, Maine, Maryland, Massachusetts, New Hampshire, New Jersey, New York, Rhode Island, and Vermont. Each state has an emissions budget, and allowance prices are uniform across the region. At least 25 percent of the auction revenues must be used to promote low carbon-energy development and offer other consumer benefits. (http://www.rggi.org/home)

RIM (Rate-payer Impact Measure)
A type of test used by utilities and regulatory agencies to evaluate the cost-effectiveness of energy efficiency programs and portfolios. It measures the impact on utility operating margins to determine if rates must increase to maintain those margins.

ROA (Return on Assets)
Net income divided by total assets. It is a metric of the profitability of a utility company relative to its total assets, or the effectiveness of a utility at using assets to generate earnings.

ROCOF (Rate of Change of Frequency)
A calculation and estimation about frequency and its rate of change that is performed by a PMU (Phasor Measurement Unit).

ROLL (Routing over Low power and Lossy networks)
The name of an IETF (Internet Engineering Task Force) working group that is defining an IPv6 routing architectural framework for these types of networks, which can be segmented as Industrial, Home, Commercial Building, and Urban. Special focus is on high reliability and connectivity for large numbers of embedded devices in a network and security and manageability concerns. (http://www.ietf.org/html.charters/roll-charter.html)

RPL (Routing Protocol for Low-Power and Lossy Networks)
An IPv6 protocol for multipoint to point traffic from devices in a low-power and/or lossy network that can transmit back and forth from a central control point.

RPS (Renewable Power *or* Portfolio Standards)
Standards that require a certain percentage of power generation must come from renewable energy sources. Standards are established for utilities on a state-by-state basis.

RS232
This is a standard for the serial communication transmission of data. It defines the methodology of connection between data terminal equipment, and data communication equipment such as a modem.

RS485
Also known as EIA/TIA 485, this standard defines the characteristics of drivers and receivers in a digital multipoint system. It is a core standard for the configuration of TCP/IP based local networks and other M2M communication linkages.

RSA
An algorithm for public key cryptography

RSE (Radiated Spurious Emission)
A radio frequency outside of a transmitter's assigned bandwidth. RSE includes harmonic, parasitic, and oscillating frequencies at the high and low bandwidth boundaries. Spurious emissions can be reduced without affecting the corresponding transmission of information.

RTEP (Regional Transmission Expansion Project)
A project that strives to identify required fixes and upgrades to an electricity transmission system to accommodate a variety of future energy scenarios, including widespread renewable sources. RTEP operates by testing the transmission system against regional and NERC national standards in order to ensure reliability and integrity.

RTLS (Real Time Locating Systems)
This standards team is focused on defining an air interface protocol that is interoperable at different frequencies for real time asset identification and management. It is Working Group 5 of the ISO/IEC JTC 1/SC 31.

RTO (Regional Transmission Organization)
These FERC (Federal Energy Regulatory Commission)-regulated entities are similar to an ISO (Independent System Operator) and have similar functions to monitor system loads and voltage profiles, operate transmission, oversee generation, and create and deploy contingency plans and emergency procedures. They are based on FERC Order 2000. ISO/RTOs also conduct spot, or Day 1, markets and day-ahead or Day 2, markets. There are four RTOs in North America: MISO, ISO-NE, PJM, and SPP.

RTOS (Realtime Operating System)
An operating system with predictable maximum execution times to fulfill the requirements of embedded systems such as those that support M2M applications. These applications fall into two classes – event response and closed loop control.

RTP (Real Time Pricing)
A pricing option or rate structure used by utilities as part of a demand response program to reduce energy use. The purpose is to encourage consumers to plan energy use during lower-priced time frames. Pricing may change based on the spot or wholesale market for day-ahead or hour-ahead energy transactions. This pricing option is currently used for the most part with C&I (Commercial and Industrial) customers, but with AMI (Advanced Metering Infrastructure) and true P2D (Prices to Devices) capabilities, it could be used with residential customers, too. It is a form of non-dispatchable resource.

RTR (Real Time Ratings)
A monitoring technology that enables a grid operator to know the true capacity of existing transmission lines in realtime instead of static assumptions that are updated only on a periodic basis.

RTU (Remote Terminal Unit)
A device that conducts sensing and transmission of information to a controller. These devices are part of SCADA (Supervisory Control And Data Acquisition) systems.

Rubee® (IEEE 1902.1 Standard for Long Wavelength Wireless Network Protocol)
A bi-directional, on-demand, peer-to-peer, radiating-transceiver protocol operating at wavelengths below 450 khz. This protocol works in harsh environments with networks of many thousands of tags and has an area range of 10 to 50 feet. RuBee is an alternative to RFID for some applications. RuBee tags may be simple identity tags with an IP address or have a four-bit processor with 500 to 5,000 bytes of static memory, optional sensors, signal processing firmware, displays, and buttons.

Run-of-the-river
A hydroelectric generation facility typically installed on rivers to capture peak or low flows or located downstream from a larger dam. These projects

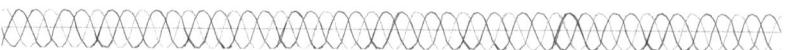

can be designed to have minimal environmental impact to the surrounding area.

Run to Failure
A maintenance practice of operating equipment until it fails, and then conducting repair or replacement of that equipment.

RUS (Rural Utilities Service)
A department formed in 1994 in the USDA (United States Department of Agriculture) that is responsible for providing electricity, telephone, water and waste services to rural areas through public-private partnerships. It offers loans and loan guarantees to finance the construction of distribution, transmission, and generation facilities, including system improvements and replacement required to furnish and improve electric service in rural areas, as well as DSM (Demand-Side Management), energy conservation programs, and on-grid and off-grid renewable energy systems. (http://www.rurdev.usda.gov/UEP_HomePage.html)

S

SAE (Society of Automotive Engineers)
A global engineering society in automotive, aerospace, and related commercial-vehicle industries focused on standards development and professional continuing education. (http://www.sae.org/)

SAE J1772 (TM) **Electric Vehicle and Plug- In Hybrid Electric Vehicle Conductive Charge Coupler**
A standard defining the connector from an electric vehicle to a charge source. It is being updated for Level 1 and Level 2 power levels (120V and 240V AC) and includes a diagnosable detection circuit and enables DC charging. A 300-to-600V DC 3-phase connector is under development for inclusion in this standard.

SAE J2293 Energy Transfer System for Electric Vehicles
A standard that establishes requirements for the transfer of electrical energy from a utility to an electric vehicle. The energy transfer system is responsible for the conversion of AC into DC electricity that can be used to charge EV storage batteries.

SAE J2836 (TM)
A standard that establishes use cases for communication between electric vehicles and the charging equipment, including diagnostic use cases about charging failures and messages to provide accurate information to the customer and/or service personnel to identify the source of the issue and assist in resolution. Other messages will inform car owners about charge status.

SAE J2847
A standard in development defining communications formats between an electric vehicle and a charge source for charge/discharge sessions.

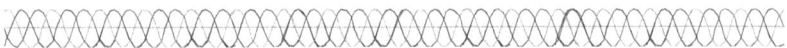

SAE J2847/2-3
A standard in development defining communications formats between an electric vehicle and a charge source for charge/ discharge sessions. Part 1 establishes energy transfer communication requirements, Part 2 outlines specifications for DC off board chargers, and Part 3 discusses linking plug-in electric vehicles with SEP (Smart Energy Profile) 2.0.

SAE J2894
A standard in development regarding power quality. It will identify AC characteristics and battery-charging parameters and recommend target values for power quality, susceptibility, and power control parameters for bi-directional energy flow.

SAE J2907
A standard in development that will describe test methods and conditions for rating performance of electric propulsion motors in electric vehicles.

SAE J2908
A standard in development that will describe test methods and conditions for rating performance of electric vehicle propulsion systems reflecting thermal and battery capabilities and limitations.

SAE J2931
A standard in development that will establish PLC-based (Powerline Carrier) communication requirements for charging equipment interfaces to HANs (Home Area Networks), HEMS (Home Energy Management Systems), or the grid.

SAFETY Act (Support Anti-Terrorism by Fostering Effective Technologies Act)
A 2002 federal law that delivers two tiers of liability protection to sellers and buyers of technologies for damages from acts of terrorism. Sellers that certify products with the SAFETY Act can be immunized from damage

claims, and buyers of certified products face no liabilities under the Act. Cyber security and monitoring applications will be impacted by this legislation.

Safety relief valve
A valve that opens rapidly for gases and slowly for liquids.

Safety valve
A valve that opens rapidly, usually used in systems containing steam and compressed air.

SAIDI (System Average Interruption Duration Index)
An IEEE (Institute of Electrical and Electronics Engineers) reliability index to track and benchmark reliability performance in utilities. It is calculated by taking the duration or time of service interruptions and dividing that by the total number of customers.

SAIFI (System Average Interruption Frequency Index)
An IEEE (Institute of Electrical and Electronics Engineers) reliability index to track and benchmark reliability performance in utilities. It is calculated by taking the number of customers impacted by an outage and dividing that by the total number of customers.

Saline water
Water that contains significant amounts of dissolved solids that is considered unsuitable for human use and may impair crop production if used in irrigation.

SALTPI (System Average Transmission Line Performance Index)
A maintenance performance metric based on the five-year average number of faults on any specific transmission line expressed in faults per hundred miles of line length per year.

SARFI (System Average Root mean square variation Frequency Index)
A power quality measurement that counts voltage sags. It is not typically included in reliability metrics.

SaveWater!® Alliance
A nonprofit Australian organization focused on water conservation and efficient water use aligned with government and water industry objectives. (www.savewater.com.au)

SB 17 (Senate Bill 17)
A California law that requires state agencies to determine the requirements for a Smart Grid deployment plan that improves overall efficiency, reliability, and cost effectiveness of electrical system operations, planning, and maintenance by July 1, 2010. Each electrical utility in the state submitted a plan to the commission for approval by July 1, 2011.

SB 412 (Senate Bill 412)
A California state law that revised the Public Utilities Code Section 379.6 to extend the SGIP program until January 1, 2016, and define eligible technologies as those distributed energy resources that the CPUC (California Public Utilities Commission) determines will achieve reductions in GHG emissions.

SCADA (Supervisory Control and Data Acquisition)
Systems used by utilities to monitor and control their generation, transmission, and distribution equipment and facilities. Components generally include I/O signal hardware, communications equipment, controllers, and software. SCADA enables utilities to stay informed about substation operations in realtime. Function examples include activation of switches, re-closers, breakers, and relays; automated fault restoration; alarms and GIS (Geographic Information System) displays; and data collection for trend analysis and load forecasting.

Scalping
Decentralized extraction of reclaimed, nonpotable water for local reuse, such as irrigation, from wastewater through use of small scale, distributed plants designed for this purpose.

SCASC (Syncophasor Conformity Assessment Steering Committee)
A branch of ICAP (IEEE Conformity Assessment Program), whose mission is to encourage adoption and industry collaboration in the expansion of PMU technology. A concurrent priority is to provide a framework for PMU vendors to check their products against standard performance criteria created by ICAP-approved laboratories.

SCC (Smart Cities and Communities)
The Smart Cities and Communities European Innovation Partnership supports the demonstration of energy, transport and information and communication technologies (ICT) in urban areas to enable innovative, integrated and efficient technologies to roll out and enter the market more easily. The Initiative supports cities and regions with projects aimed at 40% reductions of greenhouse gas emissions through sustainable use and production of energy.
(http://ec.europa.eu/energy/technology/initiatives/smart_cities_en.htm)

SCC (Standards Council of Canada)
A federal Crown corporation with the mandate to promote efficient and effective voluntary standardization. The SCC manages Canada's participation in the IEC (International Electrotechnical Commission) and ISO (International Organization for Standardization) and regional standards organizations. It also encourages the adoption and application of international standards in Canada. (http://www.scc.ca/en/web/scc-ccn)

SCC31 (Standards Coordinating Committee 31)
A committee formed by Utilimetrics for development of standards for AMR (Automated Meter Reading) and energy management for electric, gas, and

water utilities. The IEEE (Institute of Electrical and Electronic Engineers) approved a charter for SCC31, which now has several subcommittees working in areas such as security, communications interfaces, and information transfer. Utilimetrics provides financial and administrative support for SCC31, and IEEE approves and owns all finished standards. (http://www.utilimetrics.org/)

SCOE (Secure Common Operating Environment)
A Smart Grid architecture designed to address the cyber security challenges of integrating legacy systems with new technology.

SCT (Societal Cost Test)
A type of test used by utilities and regulatory agencies to evaluate the cost effectiveness of energy efficiency programs and portfolios. It measures the net economic benefits using TRC (Total Resource Cost) and externalities such as environmental benefits.

SDMS (Sensor Data Management System)
A software system that can gather data from different sensors and sensor networks and provide that data to other intra or inter domain applications.

SDO (Standards Development Organization)
Any organization that publishes specifications or standards for design, development, operation, performance, testing, and/or certification of hardware or software that is used in the Smart Grid and other business sectors. These can be international or national in scope and may focus on specific industries or technologies.

SDR (Software Defined Radio)
Software technology that enables radio transmitters and receivers to operate in different bands that may be dynamically assigned by that software, instead of hardware. It is a technology that enables cognitive radio networks.

SDWA (Safe Drinking Water Act)
Federal legislation passed in 1974 and amended in 1996. Its goal is to protect public health by ensuring that public water systems (PWSs) comply with all health-based standards via enforcement by the EPA. It specified contaminants for monitoring purposes and also required notification to consumers if maximum contaminant levels were exceeded.

SECA (Solid State Energy Conversion Alliance)
A program located within NETL (National Energy Technology Laboratory) with the objectives of lowering the cost of fuel cells by developing new materials and processes, reducing dependence on foreign oil, mitigating environmental concerns with electricity production, and providing for clean efficient power with current fuels and hydrogen. It consists of government, industry, and scientific community members to accelerate the commercial readiness of fuel cells in the 3-to-10 kW range for use in stationary, transportation, and military applications. (http://www.netl.doe.gov/technologies/coalpower/fuelcells/seca/index.html)

Security
One of two important characteristics of the reliability of the interconnected bulk power system. It is the ability of the bulk power system to withstand sudden disturbances, such as electric short circuits or unanticipated loss of system elements from credible contingencies.

Security perimeter
The boundaries of security for any grid asset or device, system, or network. It can address both physical and logical security. Perimeters help define the scope of security projects.

Sediment
A material in suspension in natural water; it includes deposits from the waters of streams, rivers, lakes or seas.

SEE Action (State Energy Efficiency Action Network)
A group of state and local governments, associations, business leaders, and non-government organizations focused on helping the USA achieve cost-effective energy efficiency by 2020 through sharing information to aid in implementation of energy efficiency policies and programs. It is co-chaired by DOE (Department of Energy) and EPA.
(http://www1.eere.energy.gov/office_eere/see_action.html)

SEEA (Southeast Energy Efficiency Alliance)
A nonprofit organization that promotes energy-efficient policies and practices through networking, program activities, and education. The SEEA is active in 11 states: Alabama, Arkansas, Florida, Georgia, Kentucky, Louisiana, Mississippi, North Carolina, South Carolina, Tennessee, and Virginia. (http://www.seealliance.org/)

SEDC (Smart Energy Demand Coalition)
A European-based industry group dedicated to making the demand side a smart and interactive part of the energy value chain and Smart Grid. This goal is accomplished through promoting demand-centered programs in the areas of demand response, HES (home energy systems), and building automation. (http://sedc-coalition.eu)

SEGIS (Solar Energy Grid Integrated System)
An initiative that includes DOE (Department of Energy), SNL (Sandia National Laboratories), industry, utilities, and universities to promote development of complete systems for integrating solar power with the electrical grid. Projects focus on grid interconnections that work with the full range of emerging solar modules, achieve reliability and resiliency, reduce costs, integrate controls for energy storage systems, and allow two-way "smart" communications between the solar power systems and the electric utilities.

SEIA (Solar Energy Industries Association)
A trade association serving the solar industry that advocates for public policies like net metering, renewable portfolio standards, and investment tax credits for a variety of supply and demand side solar technologies. (http://www.seia.org)

Self-healing grid
As defined in the ARRA Section 1306 amending EISA 2007 (Energy Independence and Security Act of 2007), a grid that is designed to detect and respond to power system disturbances, detect and isolate faults, provide location information to work crews, and automate restoration activities to minimize downtime and optimize grid performance.

Sensor
A device that delivers event-driven or time-driven data over wired or wireless communications networks.

SEP 1.0 (Smart Energy Profile 1.0)
A wireless technology specification released in June 2009 designed for the ZigBee stack to provide device support for pricing signals, demand response, load control, and metering information along with basic information such as time. The physical and media access control of the data link layer are governed by the IEEE 802.15.4 standard.

SEP 2.0 (Smart Energy Profile 2.0)
An extension and enhancement of the 1.0 profile that was released in draft form for review and comment in May 2010. It supports wired (PLC) and wireless communications and includes compatibility with new and updated IEEE and IETF standards. It is designed for IP protocol support in the ZigBee stack and includes support for new layers of PHY (physical) and MAC (Medium Access Control). It intends to harmonize with IEC (International Electrotechnical Commission) 61968 and 61850.

SEPA (Solar Electric Power Association)
A nonprofit industry association comprised of electric utilities, solar companies, and other entities interested in solar electricity. It provides customized, localized and practical advice, research and events about issues in the electric utility industry. (http://www.solarelectricpower.org/)

SEPA (Southeastern Power Administration)
A federal entity in DOE (Department of Energy) that markets electricity generated at reservoirs operated by the USA Army Corps of Engineers in the states of West Virginia, Virginia, North Carolina, South Carolina, Georgia, Florida, Alabama, Mississippi, Tennessee, and Kentucky. (http://199.44.84.82/)

Sequestration formation
As defined in the ARRA Section 1306 amending EISA 2007 (Energy Independence and Security Act of 2007), any geologic formation that can store carbon dioxide.

SERC (Southeast Reliability Corporation)
A nonprofit corporation responsible for promoting and improving the reliability, adequacy, and critical infrastructure of the bulk power supply systems in all or portions of 16 central and southeastern states. It is one of the eight NERC (North American Electric Reliability Corporation) Regional Entities. Owners, operators, and users of the bulk power system in these states cover an area of approximately 560,000 square miles. SERC is divided geographically into five sub-regions that are identified as Central, Delta, Gateway, Southeastern, and VACAR.
(http://www.serc1.org/Application/HomePageView.aspx)

Service point
An electric meter, which is the point of connection between a utility and a customer's premise wiring.

Service provider
A telecommunications carrier delivering wired or wireless voice and/or data services to customers.

Set and Forget
A product-design concept that stresses convenience for end users in instructing any energy management solution (commercial, industrial, or residential) on the desired settings that are automatically enabled until the end user makes a change.

SET Plan (Strategic Energy Technology Plan)
A strategic plan developed by the European Commission to accelerate the development and deployment of cost-effective low-carbon technologies. This plan comprises measures relating to planning, implementation, resources, and international cooperation in the field of energy technology. (http://ec.europa.eu/energy/technology/set_plan/set_plan_en.htm)

SETIS (Strategic Energy Technology Plan Information System)
An organization managed by the European Commission to be an information hub about the priority energy technologies identified by the SET Plan. (http://setis.ec.europa.eu/)

Sewer System
A separate water system that takes black and gray water from residential, commercial, and industrial consumers and either returns it to a water treatment facility or into a septic system.

SGAC (Smart Grid Architecture Committee)
A team that is part of the SGIP (Smart Grid Interoperability Panel) with responsibility for creating and refining a conceptual reference model, including lists of the standards and profiles necessary to implement the vision of the Smart Grid.

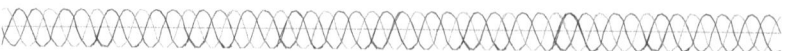

SGCSWG (Smart Grid Cyber Security Working Group)
A permanent working group for the SGIP (Smart Grid Interoperability Panel) that provides expertise needed to address matters related to cyber security for the Smart Grid. It plays a critical role in identifying the standards and architecture needed to ensure the security of the Smart Grid.

SGI (Smart Grid Ireland)
A not-for-profit industry organization focused on the benefits and opportunities in the Smart Grid sector and advocating the Smart Grid as a technology enabler. It is a member of the Global Smart Grid Federation. (www.smartgridireland.org)

SGIA (Small Generator Interconnection Agreement)
A standard agreement used to secure the interconnection of generation assets less than 20 megawatts to the grid. This agreement is part of the FERC (Federal Energy Regulatory Commission) Order 2006 and applies to ISOs (Independent System Operators), RTOs (Regional Transmission Organizations), and other transmission providers to govern interconnection.

SGIP (Small Generator Interconnection Procedure)
Procedures for the interconnection of generation assets less than 20 megawatts to the grid. These procedures are mandated by (Federal Energy Regulatory Commission) in Order No. 2006 and designed for ISOs, RTOs, and other transmission providers to modify their OATTs (Open Access Transmission Tariffs) to enable interconnection of new small generators to the grid.

SGIP (Smart Grid Interoperability Panel)
An ongoing public-private partnership partially funded by NIST (National Institute of Standards and Technology) to support its standards coordination activities and continuing evolution of the Smart Grid Reference Model. SGIP develops and reviews use cases, identifies

requirements, and proposes action plans to accelerate standards development, testing, and certification of components in the Smart Grid.

SGTCC (Smart Grid Testing and Certification Committee)
A team that is part of the SGIP (Smart Grid Interoperability Panel) with responsibility to create and maintain documentation and the organizational framework for compliance, interoperability, cyber security testing, and certification for SGIP-recommended Smart Grid standards.

SGVCA (Smart Grid Voltage Conservation Alliance)
An industry alliance comprised of Elslter, ABB, Survalent and Entergy to demonstrate end-to-end voltage conservation programs that enable operational efficiencies for utilities through distribution automation and AMI technologies. (www.sgvca.com)

SHA-1 (Secure Hash Algorithm 1)
A Federal Information Processing Standard (FIPS) issued by the National Institute of Standards and Technology (NIST) that computes a condensed representation of a message or data file. It is a technical revision of SHA or FIPS 180.

SHA-2 (Secure Hash Algorithm 2)
A Federal Information Processing Standard (FIPS)180-4 issued by the National Institute of Standards and Technology (NIST) that contains five hash functions for digital signatures, digital time-stamping, and other applications previously address by SHA-1.

SIM (Subscriber Identity Module)
A portable memory chip used in GSM (Global System for Mobile Communication^{TM}) devices, such as cell phones, to enable activation and upgrades without carrier intervention.

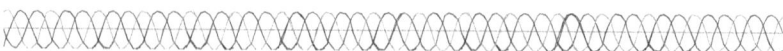

SIMON
An encryption algorithm created by the NSA in 2013 that consists of a lightweight block cipher optimized for performance in general use hardware implementations.

SIP (Session Initiation Protocol)
A signaling protocol to set up multimedia and other types of communication sessions on the Internet. It is based on RFC (Request for Comment) 3261 to RFC 3265 and is maintained within the IETF (Internet Engineering Task Force) SIPCORE working group.

SIP Forum (Session Initiation Protocol Forum)
An industry association focused on promoting the adoption and interoperability of IP communications products and services based on SIP, for use in the control of realtime multimedia communication sessions in the Internet and private networks. (http://www.sipforum.org/)

SIPS (System Integrity Protection Schemes)
A conceptual framework that proposes collection and processing of local and remote system information to the SCADA (Supervisory Control And Data Acquisition) system to preserve system stability and connectivity, and avoid major disruptions in the bulk power system. PMUs (Phasor Measurement Units) are one component that can be used to collect and relay critical data.

SISO (Simulation Interoperability Standards Organization)
An international organization dedicated to the promotion of modeling and simulation interoperability and reuse for the benefit of a broad range of M&S communities. It supports simulation interoperability standards in conjunction with IEEE and ISO/IEC. (http://www.sisostds.org/)

Situational awareness
Perception, understanding, and anticipation of the state of the power grid
system and its surrounding environment to enable effective actions in
response to changes.

SM-CG (Smart Metering Coordination Group)
A CENELEC joint advisory group defining the reference infrastructure,
identifying applicable standards, and determining functional and technical
requirements for smart meters.
(http://www.cenelec.eu/aboutcenelec/whatwedo/technologysectors/smar
tmetering.html)

Small Cell Forum
An industry association formerly known as the Femto Forum that promotes
the wide-scale adoption of small cell technologies such as femto, pico, and
metro, and microcells to improve coverage, capacity and services delivered
by mobile networks. It addresses 2G/3G/4G coverage and services within
residential, enterprise, public and rural access markets.
(http://www.smallcellforum.org/)

Smart appliance
Any electrical device that can communicate with a smart meter or an
intermediary application, such as a HEMS (Home Energy Management
System), to respond to utility signals regarding price or curtailment.
Communications may or may not be fully bi-directional.

Smart charging
Communications and charging control software that resides in any type of
EV (Electric Vehicle) and in utilities to manage the timing, pace, and extent
of charging loads from utility to vehicle and manage the load stored in the
vehicle. It can respond to fluctuations in demand on the grid, so it charges
when electricity is readily available and suspends charging when it senses

peak load times. It is generally slower than traditional charging, but preserves grid stability.

Smart energy
The use of computers, electronics, and advanced materials to make all forms of energy and fuel use more efficient. It covers the spectrum of traditional energy resources and fuels to renewable resources and alternative fuels.

Smart Energy Alliance
An industry group composed of Capgemini, Cisco Systems, GE Energy, Hewlett-Packard, Intel, and Oracle to help utilities transform their transmission and distribution operations.

Smart energy device
An energy smart consumer device that communicates via Smart Energy Profile, empowering customers to manage energy use and reduce their carbon footprint.

Smart Grid
Bi-directional electric grids and communication networks that improve the reliability, security, and efficiency of the electric system for small- to large-scale generation, transmission, distribution, storage, and consumption. It includes software and hardware applications for dynamic, integrated, and interoperable optimization of electric system operations, maintenance, and planning; distributed energy resources interconnection and integration; and feedback and controls at the consumer level. The DOE (Department of Energy) identifies seven characteristics of an electric Smart Grid:

1. Self-heals from power disturbance events
2. Enables active participation by consumers in DR (demand response)
3. Operates resiliently against physical and cyber attack

4. Provides power quality for 21st century needs
5. Accommodates all generation and storage options
6. Enables new products, services, and markets
7. Optimizes assets and operates efficiently

Smart Grid technologies can also apply to water and natural gas systems, but electricity is the only product that can be bi-directional to facilitate prosumer relationships.

Smart Grid Architecture

The standards-based logical framework for a fully interoperable and secure electrical grid enabling bi-directional electricity and communications networks.

Smart Grid Architecture Model (SGAM) Framework

A set of standards, guidelines, and regulatory templates whose goal is to ensure interoperability between two or more Smart Grid connected devices. The objective of this framework is to enable devices from different vendors to exchange information and use that information to perform the correct action for the device.

Smart Grid Australia

A nonprofit alliance and member of the Global Smart Grid Federation that is focused on establishing an electrical system based on sound technical, operational, and environmental principles to transform the infrastructure, technologies, and market in Australia. (www.smartgridaustralia.com.au)

Smart Grid Canada

A national association of public and private organizations focused on building an innovative, reliable, and cost-effective electricity delivery system that incorporates renewable energy resources and facilitates consumer interactions. It is a member of the Global Smart Grid Federation. (www.sgcanada.org)

Smart Grid Consumer Collaborative (SGCC)
A nonprofit collaborative organized to bring together utilities, suppliers, vendors, associations, and consumer advocacy groups to better understand consumer needs in the Smart Grid universe. Member companies include utilities, technology and consumer electronics companies, retailers, and consumer advocates.
(http://smartgridcc.org/)

SmartGrid GB (Smart Grid Great Britain)
An organization of trade associations, utilities, academia, and government entities to accelerate development and deployment of Smart Grid in Great Britain. (http://www.smartgridgb.org/)

Smart Grid OIR (Smart Grid Order Instituting Rule-making)
A CPUC (California Public Utilities Commission) decision that established a policy goal for California IOUs (Investor Owned Utilities) to provide consumers with access to electricity price information by the end of 2010, and that those customers with smart meters and authorized third parties have access to usage data on a near real-time basis by the end of 2011. The CPUC also stated its intent to consider and, if appropriate, adopt the NIST (National Institute of Standards and Technology) Smart Grid standards under development.

Smart Grid Standardization Mandate
A European Union mandate to develop or update a set of consistent standards that integrate digital computing and communication technologies and electrical architectures, and associated processes and Services to support interoperability of high level Smart Grid services and functionalities.
(http://ec.europa.eu/energy/gas_electricity/smartgrids/doc/2011_03_01_ mandate_m490_en.pdf)

Smart Grid Task Force
A European Union group that defined the high level services that a Smart Grid must deliver for member states. These services cover flexibility to add new users with new requirements; efficient operations; security and supply quality; plans for future investments; improved market functions and customer services; and consumer empowerment.

Smart Growth America
An advocacy group for environmentally friendly city planning. This includes walkable city routes, improved bicycle lanes, and public transit. Other projects research urban expansion, metropolitan transportation, as well as development and lobbying for resource efficiency policies at all levels of government. (www.smartgrowthamerica.org)

Smart Infrastructure
Public infrastructure systems built into a city, suburb, or rural inhabited area that autonomously manage resources and individual and aggregated assets in an optimized way through sensors, communications, and data analytics. Infrastructure systems include water, energy, buildings, and transportation, which are considered smart when they embed these technologies.

Smart inverter
A device that not only converts DC to AC electricity, but contains additional intelligence and communications capabilities to maintain "grid-awareness" and help manage load at the distribution grid. It is an important component for ubiquitous distributed generation.

Smart meter
An electric meter that is also a communications platform and a key component of the Smart Grid. It measures and records electricity use data on an interval basis specified by regulatory agencies, and can also send bi-directional messaging about demand response initiatives ranging from dynamic pricing to forms of load

control. Usage information includes more frequent measurements and time-stamped data, and is transmitted frequently to the utility. Smart meters may also provide data that assists in outage detection and restoration. It is sometimes called an advanced meter.

Smart Microgrid
A term used to differentiate the technological sophistication of a microgrid. In developed economies with well-established grids, it is presumed that any microgrid embeds Smart Grid technologies. In the developing economies with immature grid infrastructure, a microgrid may contain distributed sources of generation and energy storage but exclude the advanced communications overlay that makes a grid a Smart Grid.

Smart Networks Council
An independent operating unit of the UTC (Utilities Telecom Council) focused on addressing the evolving technology and interoperability issues for AMI (Advanced Metering Infrastructure) and DR (Demand Response) applications. One project is to build an "AMI Options Model" to help utilities ensure they are examining all available options when considering implementing a new AMI system. (http://www.smartnetworkscouncil.org/)

Smart PIG (Pipeline Insertable Gauge)
A mobile tool with multiple gauges capable of detecting leaks, corrosion evidence, and structural deviations like bulges or dents in pipelines.

Smart Plug
An intelligent wireless energy management device plugged into a wall socket and connected to a home area network or mobile device to monitor and control electrical devices plugged in to that smart plug. These devices can provide measurable power usage, centralized control capabilities, and support for energy efficiency goals by remotely controlling connected devices.

Smart technologies
Real-time, automated, and interactive technologies that optimize the physical operation of appliances and consumer devices for metering, bi-directional communication with the grid, and distribution automation, including sensors and controls. Smart devices use smart technologies.

Smart transformer
A transformer that has intelligence and communications to monitor and signal status changes, such as temperature, to schedule proactive maintenance and regulate voltage in the distribution grid.

SmartV2G
A European project to develop a Smart Electric Vehicle to Grid interface that is user friendly and easy for consumer adoption. This includes using existing data communication and vehicle to charging station protocols to broaden an intelligent urban energy supply network managed by embedded control systems. (http://www.smartv2g.eu)

Smart valve
A gas pipeline valve that can be remotely monitored and controlled to shut off as a safety measure.

SmartGridCity[TM]
An experimental Smart Grid project in Boulder, Colorado that was sponsored by Xcel to demonstrate deployment costs and benefits. For a number of reasons, the project was not successful.

SMBIOS (System Management BIOS)
This specification developed by DMTF (Distributed Management Task Force) standardizes mechanisms by which system vendors reveal management information about their products. This information is then used for generic instrumentation to deliver this data to systems that

implement CIM (Common Information Model), circumventing the slower process of remote hardware detection.

SMDS (Switched Multimegabit Data Service)
A subset of IEEE 802.6 Metropolitan Area Network (MAN) technology that is a high-speed, connectionless, public, packet-switching service offered by telecom companies. It operates at rates ranging from DS-1 or 1.5 Mbps up to DS-3 or 44.736 Mbps and uses DQDB (Distributed Queue Dual Bus) or Broadband ISDN technologies.

SMES (Superconducting Magnetic Energy Storage)
A technology that stores energy in the magnetic field created by the flow of direct current in a coil of superconducting material that has been cryogenically cooled.

SMS (Short Message Service)
A form of wireless communication that enables users to send and receive short text messages on mobile devices. It has applications in M2M communications. Also known as text messaging.

SMT (Synchronized Measurement Technologies)
Tools and systems that deliver realtime monitoring, assessing, and management of widespread grid operations. PMUs are a form of SMT.

SNIA (Storage Networking Industry Association)
A nonprofit organization that comprises some 400 global member companies in the data storage market. The SNIA focuses on the development of storage solution specifications and technologies, global standards, and storage education. The SNIA promotes the GSI (Green Storage Initiative). (http://www.snia.org/home)

Sniffing
The interception of information as it traverses a computer network and/or system. It is the computer network equivalent of eavesdropping.

SNLA (Sandia National Laboratory)
A DOE (Department of Energy) laboratory for research and development that includes the FreedomCAR program, which tests lithium-ion batteries and other energy storage technologies. This Lab is also involved in research of metal technologies to store hydrogen, SCADA security, and distributed generation. It houses the DETL (Distributed Energy Technology Center). It also includes the Sandia Water Initiative which seeks to improve the security, safety and sustainability of water infrastructure, using development of advanced technologies for reduced demand, increased supply and better decision-making tools. (http://www.sandia.gov/)

SNMP (Simple Network Management Protocol) **v3**
An internet standard protocol approved by the IESG (Internet Engineering Steering Group) for managing devices on IP networks, including remote configuration capabilities. Commonly connected devices include routers, servers, and switches among many other widely used IT assets.

SOAP (Simple Object Access Protocol)
An XML-based communications protocol that lets applications exchange information via the Internet.

Social engineering
Human interaction methods used by attackers to infiltrate organizations and networks for purposes of fraud, corporate espionage, and/or network and system disruptions.

SOFC (Solid Oxide Fuel Cell)
Fuel cells applicable for large-scale stationary power generators and co-generation facilities. Characteristics include operation between 700 and

1,000 degrees Celsius and long life. The high temperatures produce steam that can be channeled into turbines to generate more electricity, also known as co-generation.

SOL (System Operating Limit)
A NERC (National Electric Reliability Corporation) value for variables such as megawatts, amperes, or volts that satisfies the limits of necessary performance criteria for a system to ensure operation within acceptable reliability criteria.

Solar-ready home
A new home that is configured with electrical wiring and pipes to support a post-sale installation of solar water heaters and/or solar electric systems.

Solar PV (Solar Photo-Voltaic)
Solar technology that converts sunlight into electricity. PV comprises photo diodes that create DC power, which is changed to AC using inverters.

Solar thermal
Technologies that concentrate sunlight into small areas to vaporize a fluid into a gas to drive a turbine. Power towers, parabolic troughs, and parabolic dishes are three types of deployments that deliver utility-scale renewable energy for dispatchable electricity and energy storage.

Solenoid valve
Electrically operated valve that controls the flow of hot or cold water and steam.

SONET (Synchronous Optical NETwork)
An ANSI standard that defines physical interconnections for devices and network management on a fiber optic network that transmits data at speeds between 155 Mbps and 2.5 Gbps.

Specific energy
The internal energy of a substance per unit mass in Wh/kg, a metric used to describe attributes of energy storage technologies.

Specific power
The power-to-weight ratio in W/kg, a metric used to describe attributes of energy storage technologies.

SPECK
An encryption algorithm created by the NSA in 2013 for general use that consists of a lightweight block cipher optimized for performance in software implementations.

Spectrum
An ordered arrangement of radio waves for frequency, or electromagnetic waves as energy, or wavelengths as light.

SPEER (South-central Partnership for Energy Efficiency as a Resource)
An association of energy efficiency industry stakeholders committed to the accelerated adoption of advanced building systems and energy efficiency products and services in the South-central US and harmonization of state and local codes and standards. (http://www.eepartnership.org/)

Spikes
A type of power disturbance characterized by brief surges of voltage measured in milliseconds or microseconds. Spikes are often caused by lightning.

Spillway
Channels over or around dams to divert excess water.

Spinning reserve
Generation capacity that is online and consuming fuel, synchronized and ready to immediately deliver electricity to the bulk power system for

emergency events. These resources can address a demand imbalance within the first few minutes of an electric grid event. Also known as synchronous reserve.

Spoofing
The process of masquerading as another identity to gain unauthorized access to devices, systems or networks. Email, website, and IP spoofing are three common forms of deception.

SPP (Southwest Power Pool)
An RTO (Regional Transmission Organization) established in 2004 that ensures reliable supplies of power, adequate transmission infrastructure, and competitive wholesale prices of electricity. SPP oversees compliance enforcement and reliability standards development, and is also one of the eight NERC (North American Electric Reliability Corporation) Regional Entities or reliability regions. Members include 12 IOUs (Investor Owned Utilities), 9 municipal systems, and 11 generation and transmission cooperatives that serve over 5 million customers and other entities. The geographic area covers 370,000 square miles in Kansas and Oklahoma and parts of New Mexico, Texas, Louisiana, Missouri, Mississippi, and Arkansas. It is one of the 10 ISOs (Independent System Operators) or RTOs currently operating in North America. (http://www.spp.org/)

SPMS (Synchronized Phasor Measurement System)
A coordinated management and reporting system that uses PMUs (Phasor Measurement Units) to monitor power system status and dynamic transient event recording to enhance AC-DC power transmission system reliability.

Spread spectrum
A transmission technique that sends or spreads signals over a wide range of frequencies that are then converted back at the receiver to the original

bandwidth. This technique reduces interference and increases the number of simultaneous users on one frequency.

SQRA (Security, Quality, Reliability, Availability)
An extension of the power concepts for quality and reliability (PQR) that includes security and availability to reflect metrics of universal service in the electrical grid.

SRF (State Revolving Funds)
A common form of debt financing for centralized water and sanitation infrastructure.

SRNL (Savannah River National Laboratory)
The location for the Office of Environmental Management Lab that manages and operates the Center for Sustainable Groundwater and Soil Solutions. This center applies science and engineering resources and works with other agencies, universities, industry, and regulators to address DOE's (Department of Energy) priority environmental issues in cleanup activities. (http://srnl.doe.gov/)

SRS (Software Requirements Specification)
A standard found in IEEE 830-1998 that describes the guidelines to document the functions and capabilities of a software system. Also known as a Systems Requirements Specification.

SSO (Standard Setting Organization)
SSOs are entities found in virtually any industry whose primary objectives are developing, coordinating, and maintaining standards for product developers, adopters, and end-users within a given industry. They are sometimes referred to as Standards Development Organizations, or SDOs. Examples include the IEEE and IEC.

SSPC (Standing Standard Project Committee)
An acronym used by the ASHRAE (American Society of Heating, Refrigerating, and Air Conditioning Engineers) Standards Committee. An SSPC is charged with the maintenance and enhancement of an ASHRAE standard for which a group of experts needs to be available to answer questions, render interpretations, and, in general, maintain a standard after it has been promulgated. As an example, BACnet SSPC 135 is focused on building automation and controls.

SSSC (Static Synchronous Series Compensator)
A FACTS (Flexible AC Transmission System) technology that controls line impedances in high voltage AC transmission lines.

Standards
A formal document that contains technical specifications or criteria for consistent application as rules or guidelines to ensure that materials, products, processes, and services deliver expected minimum performance levels.

Standby demand
The amount of energy to be used as a backup for an outage of a customer's primary power source.

STATCOM (Static Compensators)
Components that help reduce transmission losses through control of reactive flow and provision of local reactive support. It is a FACTS (Flexible AC Transmission System) device.

Static IP
A schema used in closed networks to assign an IP address on each device in that network.

Statistical models
A type of forecast used in wind power production forecasts. Statistical models are used in combination with NWP models.

Stirling engines
External combustion engines that are sealed systems with an inert working fluid or gas, usually either helium or hydrogen. The working gas is compressed in one region of the engine and transferred to another region, where it is expanded and then returned to the first region for recompression, moving back and forth in a closed cycle. They are generally found in small sizes (1 to 25 kW) and are a possible energy source for distributed generation.

Storage-ready home
A new home that is configured with electrical wiring to support a home battery solution.

Submetering
A practice of installing additional meters behind a master meter to accurately assess unit-based energy use in multi-family residential buildings and multi-tenant commercial spaces for billing purposes.

Substation
A utility location where power is reduced from transmission high voltage to lower voltage used by customers. These facilities contain switches, capacitors, transformers, and additional equipment to manage the flow of electricity.

Substation automation
Provisioning substations with additional communications capabilities with IEDs (Intelligent Electronic Devices) to deliver realtime monitoring and remote control of equipment to improve reliability, security, and management of assets and reduce operating costs.

Substation gateway
A device that performs entry/exit functions between a substation and the distribution grid's communications network. Protocol conversion may or may not be required, but some processing generally occurs to manage traffic on the network.

Subsynchronous resonance
A class of electrical grid conditions related to oscillations at frequencies below the typical grid base frequency (in North America it is 60 Hz) that are exchanged between the power grid and a turbine generator. These usually occur in the 5-10 MHz range on AC grids.

Summer peak
Average demand from 2 to 5 PM for three successive business days with the highest average maximum temperatures. It is one type of measure of peak capacity that is used to develop a value for peak load reduction.

Superconductivity
A material property to carry electricity without resistance, so less electricity is lost while conducted through the material. The materials must be cooled to achieve superconductivity. The cooling temperatures are around minus 320 degrees Fahrenheit and are achieved with inert liquid nitrogen.

Super grid
A future grid that will provide links between today's regional grid layouts and planned renewable energy generators to transmit electricity to any region where needed. The vision for this grid is also to eliminate congestion problems and balance loads from intermittent energy sources across regions. Also known as a mega grid or national grid.

Supplemental reserve
Energy that is stored or additional generation assets that can be brought online to provide backup to spinning and non-spinning reserves. Also known as replacement reserve.

Surface water
Water located on the surface of the earth, such as streams, rivers, lakes or reservoirs.

Surge
An increase in pressure and flow in a water system that can occur when pumps are switched or when valves and hydrants are operated, resulting in a deterioration of water quality.

Sustainable Energy for All
A UN initiative to provide global universal access to modern energy services; double the global rate of improvement in energy efficiency; and double the share of renewable energy in the global energy mix by 2030. www.sustainableenergyforall.org

Sustainable Infrastructure Initiative
An EPA initiative that promotes investment in the USA water infrastructure. It also identifies best practices in water utilities and encourages adoption of these practices to address water management challenges. (http://water.epa.gov/infrastructure/sustain/index.cfm)

SVC (Static VAR Compensator)
A FACTS (Flexible AC Transmission System) device that delivers high-speed voltage support and increases transmission capacity and efficiency on AC lines. SVCs are useful to control and respond to changes in grid conditions, and can accommodate wind power and other forms of remote generation.

SVR (System Voltage Reduction)
An emergency control action to manage grid reliability when electricity demand exceeds supply. It is also known as voltage reduction and is typically in the range of a 3-to-5 percent decrease in overall system voltage. It is done in advance of rotating blackouts.

SWCC (Small Wind Certification Council)
An organization that certifies small wind turbines that meet or exceed the American Wind Energy Association's (AWEA) standards for performance and safety to assist buyers in determining product quality. (http://www.smallwindcertification.org/)

SWEEP (Southwest Energy Efficiency Project)
A nonprofit organization that promotes greater energy efficiency in a six-state region consisting of Arizona, Colorado, Nevada, New Mexico, Utah, and Wyoming. Its current programs target energy efficiency in the industrial and transportation sectors. SWEEP collaborates with utilities, state agencies, local governments, environmental groups, universities, private businesses, and other energy specialists. (http://www.swenergy.org/)

SWPA (Southwestern Power Administration)
A federal agency within DOE (Department of Energy) responsible for marketing the hydroelectric power produced at 24 USA Army Corps of Engineers multipurpose dams in Arkansas, Kansas, Louisiana, Missouri, Oklahoma, and Texas. (http://www.swpa.gov/)

Synchrophasor
A precise grid measurement or synchronized phasor that delivers realtime data about the power system. The information is obtained from monitors called PMUs (Phasor Measurement Units). Aggregating this time-stamped or synchronized data is useful to delivering a comprehensive view of an

interconnect system. These measurements are used for wide area management of grid operations.

Synchrophasor Class
A functional designation that describes the measurement or protection data collection capabilities of synchrophasors.

Synchrophasor Registry
(*See* PMU Registry)

System coincident peak
Maximum MW demand at a utility for a single hour in a year. It is one type of measure of peak capacity that is sometimes used for long-term generation planning purposes.

System engineering
The practice of gathering requirements and then optimizing a specific system design to those requirements and budget.

System of systems
A meta concept that enables consideration of the electrical grid as a network of systems functioning together for a common purpose. It also influences an emerging practice of system engineering for solving large, complex problems.

System operator
An individual at a control center responsible for realtime monitor and control of the electric system. Balancing authorities, transmission operators, generator operators, and reliability coordinators are all considered system operators. The system operator is responsible for initiating DR (Demand Response) events.

System peak response transmission tariff
A rate structure in which interval-metered customers reduce load during coincident peaks as a way of reducing transmission charges. It is a form of non-dispatchable resource.

T

T & D (Transmission and Distribution)
Two important functions performed by utilities in which transmission is the movement of bulk power across high-voltage facilities to substations, and distribution is the movement of lower-voltage power from substations to end users.

TAC (Technological Advisory Council)
An advisory council within the FCC (Federal Communications Commission) that provides technical advice to the FCC and makes recommendations to the FCC Chairman, Chief Technologist, and Chief of the FCC Office of Engineering and Technology, or other designated federal resources. The focus is on technological and technical issues in communications, and may include spectrum policy, broadband technology and deployment, communications technology that enhances and supports public safety, Internet security, and emerging systems such as the Smart Grid.

TAHI (The Application Home Initiative)
An industry association that promotes interoperability and applications using broadband networks for connected homes, grids, and cities. (http://www.theapplicationhome.com/)

TASC (The Alliance for Solar Choice)
A trade group formed by the leading rooftop solar companies to promote the widespread residential use of solar panels. A primary goal of this group is ensuring the continuation of net metering which is the practice of compensating solar consumers through cash or credits for directing energy from their solar panels back to the grid.

Tamper detection
A smart meter ability to alert the utility if there is unauthorized access detected on a meter.

Tariff
A utility's published rate schedules and general terms and conditions for supplying electricity to customers.

TC57
See IEC Technical Committee 57.

TC 224
The ISO Technical Committee responsible for standardization of a framework for the definition and measurement of service activities relating to drinking water supply systems and wastewater systems.

TC M2M (Technical Committee Machine to Machine)
ETSI's (European Telecommunications Standards Institute) working group focused on developing and maintaining standards to support an M2M IP connection from connectivity to services.

TCP/IP (Transmission Control Protocol/Internet Protocol)
A low-level protocol used in Ethernet networks and by higher-level protocols to transport data. It performs error-checking functions to improve reliability of data and reassembles packets at their destination.

TCR (Total Capital Requirement)
A utility economic analysis for funding projects. It can be done in current or constant dollars. Also known as All-in Costs.

TCSC (Thyristor Controlled Series Compensator)
A FACTS (Flexible AC Transmission System) that controls line impedances to improve voltage stability on high voltage AC transmission lines.

TDMA (Time Division Multiple Access)
A communications standard that divides calls into time slots on a designated channel. This method allows a single channel to carry many calls

at once. Commonly used in GSM (Global System for Mobile Communication™) 2G networks, it is not typically found in 3G networks.

TDS (Total Dissolved Solids)
A quality metric for organic and inorganic particulates in water.

TDSP (Transmission/Distribution Service Provider)
An entity that owns or operates for compensation the equipment or facilities to transmit and/or distribute electricity in Texas. TDSPs are regulated by the PUCT (Public Utility Commission of Texas) and must provide non-discriminatory access to the grid.

TDU (Transmission Distribution Utility)
An entity in the Texas deregulated market that is prohibited from owning generation or taking possession of electricity.

Telecommunications
The transmission of analog or digital signals or data across a distance by wired or wireless means.

Telemetering
The process by which substation and generating station metrics are transmitted to the control center and may trigger commands back from the control center. Metrics can include instantaneous voltage, current, energy, status, and alarm information. It is also a process to remotely acquire meter readings through communication devices.

TES (Thermal Energy Storage)
Technology that can efficiently enhance the productivity of cooling, heating, and refrigeration systems. Two types of systems are typically deployed. The first system uses a material that changes phase, usually water and ice. The second type changes the temperature of a material, usually water. Also known as cool storage.

TESA (Texas Energy Storage Alliance)
An industry coalition that provides advocacy and education on energy storage market opportunities and barriers to offering open and fair environments in Texas. (http://www.texasenergystorage.org/)

TG4e (IEEE 802.15.4 Task Group 4e)
An IEEE (Institute of Electrical and Electronics Engineers) group that is adding functionality to the 802.15.4-2006 to support applications in factory and process automation, asset tracking, C&I (Commercial and Industrial) sensor control that includes building automation, and neighborhood area networks. (http://www.ieee802.org/15/pub/TG4e.html)

TG4g-SUN (IEEE 802.15.4 Task Group 4g – Smart Utility Networks)
An IEEE (Institute of Electrical and Electronics Engineers) group tasked to create an amendment to the IEEE 802.15.4 to facilitate very large-scale process control applications, such as the utility Smart Grid telemetry network. The special characteristics of a Smart Grid, large, geographically diverse, with large numbers of fixed end points combined with the need for minimal infrastructure, will be addressed by this group. (http://www.ieee802.org/15/pub/TG4g.html)

The Open Group
See Open Group.

Thermal distillation
A desalination process that boils water to steam and condense in a separate reservoir, removing contaminants with higher boiling points than water.

Thermal load
Any amount of excess thermal energy or heat that is transferred or exchanged for cooling purposes. Waste heat commonly uses fluids like water to discharge thermal loads.

Thermal pollution
An increase in ambient water temperature through discharges from water treatment facilities, industrial facilities, or power plants that degrades the local water quality or environment.

Thermoelectric generation
Generation plant technologies that convert thermal energy into electrical energy and require significant amounts of water to cool and condense steam after exiting a turbine. There are two conversion methods – the Brayton cycle and the Rankine cycle. There are three common systems for cooling: once-through, wet recirculating, and dry cooling.

Third-party aggregator
An entity that signs up customers for special programs, such as energy efficiency, with the knowledge and support of the utility that supplies energy to those customers. Also known as a third-party affiliate.

Threat
The potential actions that can compromise the security of data, devices, systems, grids, or networks.

Threat model
A methodology that lists vulnerabilities, threats, and mitigations to proactively address security issues.

TIA (Telecommunications Industry Association)
An industry association of manufacturers of global information and communications technology (ICT) equipment and networks that promote standards development, policies, and business for its members. (http://www.tianow.org/)

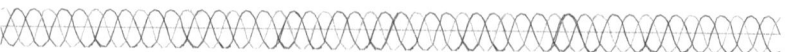

TIA 102
A land mobiles radio communications standard, which details the interface between mobile, portable, and base station radios. This standard has many parts and is frequently updated as land mobile radio technology advances.

TIA 4940
The Smart Device Communications Reference Architecture standard that defines M2M communications requirements and interoperability of smart devices for both wired and wireless transport layers and take advantage of IP-enabled applications. It applies to smart metering and building automation applications as well as other IoT applications.

TIA TR-50
An M2M protocol series to facilitate the wired and wireless interconnection of devices from a multitude of IED (intelligent electronic device) manufacturers. Also included is a vendor agnostic API to ensure development of new products that are interoperable with existing devices from other vendors.

Tie line
A circuit connecting two balancing authority areas.

Time of use rates
Price schedules used to deliver pricing feedback to consumers to encourage electricity consumption changes. Also call TOU rates.

Time-shifting consumption
The desired result of real-time pricing, which works to meld peak supply with peak demand.

Time synchronized
The designation of an instant on a time scale that is accurately calibrated with all devices on a network. It refers to monitor and control functions that are important for many grid-critical operations, including NASPInet and generation, transmission, and distribution at national, regional, or local levels.

TinyOS (Tiny Operating System)
An open-source operating system designed for wireless embedded sensor networks. It features a component-based architecture that enables rapid innovation and implementation while minimizing code size as required by the severe memory constraints inherent in sensor networks.

Title 24 (Energy Efficiency Standards for Residential and Nonresidential Buildings)
Part 6 of this California Code of Regulations mandate was set up in 1978 to develop energy-efficiency standards for California buildings. These standards are updated periodically and the new ones took effect on 1/1/2010. (http://www.energy.ca.gov/title24/)

TLPI (Transmission Line Performance Index)
The five-year average of the number of faults on any specific transmission line expressed in faults per hundred miles of line length per year.

TLR (Transmission Loading Relief)
Procedures to prevent or manage potential or actual system operating limit or interconnection reliability operating limit violations of Reliability standards – particularly IRO-006-3.

TMR (Trunk Mobile Radio)
Widely used in dispatch and fleet applications, such as police, fire, and other emergency vehicles and taxi and utility fleets. Also known as Specialized Mobile Radio.

TOGAF (The Open Group Architecture Framework)
An architecture framework for use in developing information systems architecture that was created by The Open Group's Architecture Forum. The current TOGAF Version 9 describes methodology and supporting resources to develop enterprise architectures.

Torrefaction
A thermo-chemical treatment of biomass ranging from 200 to 340 degrees Celsius with the objective of removing water and volatile compounds, and increasing energy content by 30% per unit of mass. The resulting material is sometimes called biocoal.

TOU (Time of Use)
A rate structure with different unit prices for electricity use in a 24-hour time frame, generally to encourage use during periods of lower demand. This phrase applies to a time of use price, rate, or tariff and is a dynamic price scheme typically used with non-dispatchable demand response programs. Also known as time of day pricing.

Total internal demand
This load amount equals the sum of the Internal Demand and the Standby Demand.

TR 45 (Technical Report 45)
This is a set of open standards from TIA (Telecommunications Industry Association) for CDMA2000 spread spectrum and high rate packet data systems. These standards are specifically applied to mobile and point to point networks.

TR 51 (Technical Report 51)
A TIA (Telecommunications Industry Association) standard under development to provide with a multipart wireless mesh network topology. This can be applied to smart meter networks.

TR 69 (Technical Report 69)
A Broadband Forum protocol for secure communications between equipment at a customer location and an auto-configuration server that exists within a broadband provider's network to manage the configuration of the CPE (customer premise equipment).

TR 98 (Technical Report 98)
A Broadband Forum data model for Internet gateway devices that use TR 69 protocol for communications.

TR 143 (Technical Report 143)
A Broadband Forum document that defines data model objects for network performance testing for TR 69.

Transactive Energy
A software-defined grid managed via market-based incentives to ensure grid reliability and resiliency. This is done with software applications that use economic signals and operational information to coordinate and manage devices' production and/or consumption of electricity in the grid. Transactive energy describes the convergence of technologies, policies, and financial drivers in an active prosumer market- where prosumers are buildings, EVs, microgrids, VPPs or other assets.

Transformer
An electrical device that changes voltage in direct proportion to current (higher or lower) for transmission and distribution of electricity. Transformers "step up" or "step down" voltages.

Transient Stability
Any situation that involves major disturbances, such as loss of generation, faults, and sudden load changes. Following a disturbance, synchronous machine frequencies undergo transient deviations from synchronous frequency (60 Hz in the U.S) and machine power angles change. A transient

stability study can determine whether or not machines will return to synchronous frequency with new steady-state power angles after a disturbance.

Transmission Grid
The hardware and software systems that manage and deliver high voltage electricity from generation sources to distribution grids, including power lines and substations. Voltages may be AC (alternating current), or DC (direct current).

Transmission level
The voltage range for transmission of electricity; for long distance transmission, it is generally 69 KV or higher.

Transmission operator
The entity responsible for the reliability of its transmission system directing the operations of the transmission facilities.

Transmission planner
One of the regional operational functions in the bulk power system to ensure a reliably interconnected bulk power system.

TRC (Total Resource Cost)
A type of test to evaluate the cost effectiveness of energy efficiency programs and portfolios. It compares the direct costs that both utilities and ratepayers pay to the net energy efficiency benefits. It is a test that is used by regulatory agencies to determine cost-effective energy-efficiency measures.

TRE (Texas Regional Entity)
A functionally independent division of ERCOT (Electric Reliability Council of Texas) that is authorized by NERC (North American Electric Reliability Corporation) to develop, monitor, assess, and enforce compliance with

NERC reliability standards within the geographic boundaries of the ERCOT region. It is one of the eight NERC Regional Entities or reliability regions. It is also authorized by the PUCT (Public Utility Commission of Texas) and is permitted by the NERC to investigate compliance with the ERCOT protocols and operating guides, working with Commission staff regarding any potential protocol violations. (http://www.ercot.com/mktrules/compliance/tre/index)

TRL (Technology Readiness Level)
A term used within USA governmental agencies to enable tracing progression of a technology from applied research to commercialization. There are nine levels of progression.

TRM (Transmission Reliability Margin)
The amount of transmission transfer capability necessary to provide reasonable assurance that the interconnected transmission grid will be secure. TRM accounts for the inherent uncertainty in system conditions and the need for operating flexibility to ensure reliable system operation as system conditions change.

Trojan horse
A computer virus that masquerades as an authorized or beneficial application. Its objective is to steal valuable information and take control of infected computers or servers.

TS 103 908 (Technical Specification 103 908)
An OSGP (Open Smart Grid Protocol) specification originating from ESNA and published by ETSI for Narrow Band Power Line Channel for Smart Metering Applications. It defines a high-performance narrow band powerline channel for control networking that can be used with multiple Smart Grid devices.

TSC (Transient Stability Control)
Systems that monitor and analyze transmission conditions in realtime to optimize transmission stability and include actions to preserve stability and prevent failures.

TSO (Transmission System Operator)
An entity responsible for the bulk transmission of electricity on high voltage grids, and grid access to generators, traders, suppliers, distributors and directly connected customers. Other responsibilities and roles vary by country.

TTA (Telecommunications Technology Association)
An industry organization focused on developing technical standards for telecommunications and testing and certification processes. (http://www.tta.or.kr/English/new/about/messagefrom.jsp)

TTC (Telecommunication Technology Committee)
A Japanese-based organization that establishes protocols and standards for telecommunications networks and terminal equipment and promotes those standards. (http://www.ttc.or.jp/e/)

TVIC (Tennessee Valley Industrial Council)
A trade group comprised of representatives from industrial interests that purchase power from the Tennessee Valley Authority. The committee's main objective is to negotiate favorable power prices from TVA for industrial production to continue in the area.

TVPPA (Tennessee Valley Public Power Association)
A nonprofit organization that serves as an advocate and business service provider for consumer-owned local power distributors within the Tennessee Valley Authority. TVPPA operates in seven different states (Alabama, Georgia, Tennessee, Mississippi, Kentucky, North Carolina, and

Virginia) which totals to approximately nine million customers. (www.tvppa.com)

Two-way power flow
The bi-directional flow of electricity on the grid that enables H2G (Home to Grid), V2G (Vehicle to Grid), B2G (Building to Grid) and other DER (Distributed Energy Resource) scenarios to be supported.

U

UCA (Utility Communications Architecture®)
A suite of existing technologies and standards, including Ethernet, TCP/IP or ISO-OSI, and an MMS (Manufacturing Message Specification) application layer. It is the result of an organized initiative by multiple utilities (electric, gas, and water) and EPRI (Electric Power Research Institute) to standardize communications among the various components of their operations.

UCAIug (Utility Communications Architecture International Users Group)
A not-for-profit corporation of utility users and vendors promoting the integration and interoperability of electric/gas/water utility systems with international standards-based technology. It has three affiliated member groups: IEC61850ug, CIMug, and OSGug. The UCAIug does not write standards but works closely with those bodies that have primary responsibility for the completion of standards, including IEC TC57. It is also working with DRRC (Demand Response Research Center) on the OpenADR specification. (http://ucaiug.org/default.aspx)

UCLA WIN Smartgrid (University of California, Los Angeles Wireless Internet SmartGrid)
A network platform technology (wireless, RFID (Radio Frequency Identification), and integrated sensors) developed by UCLA that delivers wireless monitor and control of devices to a Web-based service, and conducts dynamic, real-time communications with electric utilities. (http://winmec.ucla.edu/smartgrid/2009-06/)

UCOWR (Universities Council on Water Resources)
An international association of 90 member universities and organizations that provides information and education on water issues and promotes research on water topics. Its activities include an annual conference and publication of a journal on water topics and news. (http://www.ucowr.org/)

UCT (Utility Cost Test)
A type of test to evaluate the cost effectiveness of energy-efficiency programs and portfolios, it is also known as the program administrator cost test. It measures the total utility costs against changes in revenue that result from energy- efficiency programs. This test is used by utilities that want to minimize their life cycle revenue requirements.

UDP/IP
A low-level IP protocol for Ethernet networks. Unlike TCP/IP, it does not provide error control. It does not require handshakes or block data like TCP/IP, making it useful for certain realtime data communications needs.

UFW (Unaccounted-For-Water)
A term categorizing water that is lost in water systems, used for firefighting, or water used in treatment processes.

UHVDC (Ultra High Voltage Direct Current)
Electricity that is transported in the transmission grid at over 800kV.

UIC (Underground Injection Control)
A program to safeguard underground supplies of drinking water established by the SDWA (Safe Drinking Water Act) that may be affected by the process of introducing fluids, slurries or gases into subterranean formations for storage. It ensures that injection wells do not compromise current and future underground sources of drinking water.

UL (Underwriters Laboratories®)
An independent product safety certification organization that tests products and writes standards for safety.

UL 1741
A standard that covers inverters, converters, charge controllers, and interconnection system equipment intended for use in stand-alone or DER

power systems. It also impacts energy storage.
(http://ulstandardsinfonet.ul.com/scopes/scopes.asp?fn=1741.html)

UL 1778
A standard for movable, stationary, fixed, and built-in uninterruptible
power systems up to 600V AC in distribution grids.
(http://ulstandardsinfonet.ul.com/scopes/scopes.asp?fn=1778.html)

UL 2202
A standard for EV charging-system equipment that may be located on or off
the vehicle. Equipment is intended to be installed in accordance with the
National Electrical Code, NFPA (National Fire Protection Agency) 70.

UL 2231
A standard defining the safety protections that must be in place regarding
EV supply circuits to reduce the risk of electric shock to the user from
accessible parts in grounded or isolated circuits for charging electric
vehicles. These circuits are external to or on board the vehicle.

UL 2251
A standard for connectors, such as plugs, receptacles, and couplers, rated
up to 600 volts AC or DC in accordance with National Electrical Code NFPA
(National Fire Protection Agency) 70 for either indoor or outdoor
nonhazardous locations.

UL 2594
A standard that addresses EV (Electric Vehicle) cord sets and EV charging
stations rated a maximum of 250 V AC and intended to provide power to an
electric vehicle with an on-board charging unit. Cord sets may be portable
or stationary and for indoor or outdoor use. Charging stations may be
either movable or permanent charging stations for indoor or outdoor use.
Intended for use in accordance with the National Electrical Code NFPA
(National Fire Protection Agency) 70.

Ultracapacitor
An energy storage device that increases the energy density of a regular capacitor through use of carbon or highly porous electrode materials. Ultracapacitors have lower energy density than lead-acid batteries, but can be cycled tens of thousands of times and are much more powerful than batteries with regard to fast charge and discharge capabilities. Also known as a supercapacitor.

UMTS (Universal Mobile Telecommunications System)
A 3G broadband and packet-based mobile telecommunications technology that is positioned as the successor to GSM (Global System for Mobile Communication). It uses W-CDMA signaling technology and achieves rates ranging from 384 kbps to 2 Mbps. It is sometimes known as 3GSM, although it is incompatible with GSM.

UMTS 900 (Universal Mobile Telecommunications System 900)
The deployment of UMTS technology in the 900-MHz GSM band. It offers improved indoor coverage in urban areas and use of larger cell radii in rural areas, reducing deployment costs.

UMTS Forum (Universal Mobile Telecommunications System Forum)
An international industry association focused on 3G wireless technologies and its evolution. It promotes a common vision and conducts studies on markets and trends, spectrum and regulation, and technology issues. It is also working on a road map of 3G/UMTS evolution that conforms with 3GPP. Members include network operators, regulators, and the manufacturers of network infrastructure and terminal equipment. (http://www.umts-forum.org/)

Undervoltage
A temporary reduction in voltage that lasts longer than a brief period of time like a second. When it is less than this timeframe, it is typically a voltage sag.

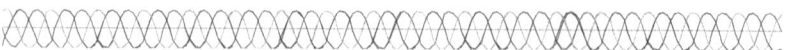

UNECE (United Nations Economic Commission for Europe)
The parent organization of UN/CEFACT (Centre for Trade Facilitation and Electronic Business), it supports activities to improve the effective exchange of products and services. Along with OASIS, it sponsored development of ebXML. (http://www.unece.org/cefact/about.htm)

UNEP (United Nations Environment Programme)
A UN-based organization that is coordinating the annual World Water Day (WWD) campaign on behalf of UN-Water to draw global attention to the importance of fresh water and the need for sustainable management of water resources. It has liaison status with seven ISO standards-development groups, including ISO/TC 147.

Unicast
Transmitting a signal or data to a single destination. It is the opposite of multicast communications.

UPFC (Unified Power Flow Controller)
A FACTS (Flexible AC Transmission System) technology that modifies and controls voltage, line impedance, and phase angle for AC power.

UPN (Utility Planning Network)
An organization with three membership-based peer groups devoted to networking, education, and advocacy. The Global AMI (Advanced Metering Infrastructure) Utility Peer Group includes utilities involved with advanced metering. The two other peer groups are for utility business and information technology customer care professionals and energy managers in large C&I (Commercial and Industrial) businesses. (http://www.utiliplan.com/)

UPS (Uninterruptible Power Supply)
Equipment that provides emergency power on short notice.

Uptake
The amount of atmospheric carbon that individual organisms or biological environments can absorb and break down in a given time period. On a macro level, the ocean's carbon uptake is approximately 2.2 petagrams of carbon per year. On the micro level, each individual photosynthetic organism has its own uptake value.

Urban Land Institute
A nonprofit research and education organization representing land use and real estate development interests covering both the public and private sectors. Smart Grid relevant goals include the advancement of sustainable building technologies, innovation in the urban development space, and the promotion of best practices in environmental construction and development. (http://www.uli.org)

URCI (Universal Remote Controlled Interrupters)
A distribution-class protective device that consists of a vacuum switch coupled with intelligent relays capable of identifying and isolating faults automatically.

USB (Utilities Standards Board)
A consortium of six North American-based utilities (AEP, Dominion, Duke Energy, Exelon, Hydro One, and Pepco Holdings, Inc) working to standardize Smart Grid system functionality and communications interfaces for complete interoperability between AMI (Advanced Metering Infrastructure) systems and utilities' operations. Current work includes meter/head-end event codes, remote connect/disconnect, and outage management. (http://www.utilitystandardsboard.com/index.html)

USCHPA (United States Clean Heat & Power Association)
A nonprofit trade association that advocates, networks, educates, and distributes marketing information to CHP (Combined Heat and Power) and distributed generation companies. It also participates in federal agency and state programs to promote CHP and clean distributed energy, such as DOE

(Department of Energy), EPA (Environmental Protection Agency), and national laboratories, including NREL National Renewable Energy Laboratory), Oak Ridge, Pacific Northwest, and Lawrence Berkeley labs. (http://www.uschpa.org/i4a/pages/index.cfm?pageid=1)

USDA (United States Department of Agriculture)
A federal agency focused on food, agriculture, and natural resources. It has oversight of some public and private lands, and provides grants, loans and loan guarantees to small communities for water supply and sanitation along with technical assistance and training. (http://www.usda.gov/wps/portal/usda/usdahome)

USDW (Underground Sources of Drinking Water)
Various sources of underground drinking water like aquifers, providing a significant percentage of water to USA public water utilities.

USEA (United States Energy Association)
An association that represents the USA energy sector on domestic and global energy issues. Its members include public and private energy-related organizations, corporations, and government agencies. The USEA is the USA Member Committee of the World Energy Council. The USEA sponsors the Energy Partnership Program in conjunction with the USAID and DOE (Department of Energy). (http://usea.org/)

USGBC (United States Green Building Council)
A nonprofit organization that certifies sustainable businesses, homes, hospitals, schools, and neighborhoods. The USGBC is dedicated to expanding green building practices and education and its LEED® program. (http://www.usgbc.org/)

U-SNAP (Utility Smart Network Access Port)
A specification that is under development by companies in the utility industry intended to be a low-cost protocol-agnostic, interoperable

communications card standard for connecting HAN devices to smart meters. It will be modeled on the USB standard for attaching hardware devices in a computer and includes a standard connector, PCB interface, and serial interface enabling consumer products to support a variety of communication protocols.

U-SNAP Alliance (Utility Smart Network Access Port Alliance)
A new industry association developing a standard for connecting energy-aware consumer products with smart meters. In addition to the standard, it plans to establish testing and certification procedures for product conformance. (http://www.usnap.org/)

UTC (Coordinated Universal Time)
A French acronym for a standard time scale equivalent to mean solar time at the prime meridian (0° longitude), popularly known as Greenwich Mean Time.

UTC (Utilities Telecom Council)
A global trade association dedicated to creating a favorable business, regulatory, and technological environment for companies that own, manage, or provide critical telecommunications systems in support of their core business. Members include over 500 utilities and technology companies and a dozen of the industry's largest energy, water, and other critical infrastructure trade associations. It is affiliated with the UPLC, UTC Canada, and Smart Networks Council. (http://www.utc.org/)

Utilimetrics
An industry alliance focused on the development and application of advanced metering and communications services. Over 1,400 members represent gas, water, and electric utilities and manufacturers in these industries. Members develop and implement automated resource-management technologies and participate in standardization and regulatory activities. It provides administrative and financial support to

SCC31 (Standards Coordinating Committee 31). It was formerly known as the AMRA (Automatic Meter Reading Association). (www.utilimetrics.org)

UtiliSite Council
An organization of utilities that makes utility infrastructure, including fiber networks and facilities, available to wireless service providers for wireless and backhaul communications needs. Members include Duke Energy and Portland General Electric Co. (http://www.utilisite.org/)

UtilityAMI WG (UtilityAMI Working Group)
A group that provides a forum for utilities to define serviceability and security and interoperability guidelines for AMI (Advanced Metering Infrastructure) and DR (Demand Response) infrastructures from a utility and an energy service provider perspective. The AMI Enterprise task force, AMI Network task force, and OpenHAN task force are situated within this group, which reports to the OpenSG subcommittee within the UCAIug (Utility Communications Architecture International Users Group). (http://osgug.ucaiug.org/utilityami/default.aspx)

UWB (Ultra Wide Band)
A wireless frequency channel allocation that ranges from 3.1 GHz to 10.6 GHz. Each channel bandwidth can be more than 500 MHz to enable devices to function with a very high data throughput as long as they are within close proximity.

UWIG (Utility Wind Integration Group)
An industry association to accelerate the development and application of good engineering and operational practices supporting the appropriate integration of wind power into the electric system. Formerly known as the Utility Wind Interest Group, its members include utilities, transmission system operators, and corporate, government, and academic organizations. (http://www.uwig.org/)

V

V2G (Vehicle to Grid)
The concept of using stored energy from PHEVs (Plugin Hybrid Electric Vehicles) and BEVs (Battery Electric Vehicles) as dynamic electrical storage. Ideally, V2G would use off-peak generation of electricity to charge batteries, which increases the asset utilization of utilities, and then sell power back to the grid during peak hours as needed. Some in the industry suggest that this should be G2V (Grid to Vehicle) to orient the thinking to the strategic role vehicles can play as distributed energy storage units for the larger grid.

V2G-enabled
EVs (Electric Vehicles) that are capable of sending power back to the grid. The EV and the grid it is plugged into must have the following: communications, price signals, charge interruption features, and billing support.

V2H (Vehicle to Home)
The concept of using stored energy from PHEVs (Plugin Hybrid Electric Vehicles) and BEVs (Battery Electric Vehicles) to connect to homes for charging and discharging.

V2I (Vehicle to Infrastructure)
A wireless exchange of safety and operational data between vehicles and highway infrastructure for collision avoidance and other safety and environmental benefits.

V2M (Vehicle to Microgrid)
A variation of the concept of using stored energy from EVs (Electric Vehicles) as dynamic electrical storage that is confined to a microgrid.

V2V (Vehicle to Vehicle)
A vehicular wireless communications ability occurring between vehicles for applications such as collision avoidance and other proactive safety measures to improve situational awareness and optimize driving safety.

Valve
A manual or automatic fluid-controlling element in a water or wastewater distribution system. Valves can start, stop, regulate or direct liquid flows; prevent backflow; and regulate pressure within the system. Valve types include gate, ball, butterfly, and globe valves.

VAR (Volt Amperes Reactive)
The measurement of reactive power.

Variable costs
Some of the costs of providing electricity to ratepayers. It includes fuel expenses for generating facilities and purchased power.

VEE (Validation, Estimation, and Editing)
Procedures that take raw meter data and perform validation and, as necessary, edit and estimate corrupt or missing data, to create validated data. Each utility may have distinct processes developed to manage this set of tasks. Before the introduction of smart meters, it was a way to develop monthly bills even in the absence of monthly meter reads.

VEN (Virtual End Node)
A term found in the Energy Interoperation specification to describe device or asset roles. It helps build a model for electricity generating devices to communicate with a a larger grid network (leveraging Virtual Top Nodes or VTNs) and enables the vision of Transactive Energy markets.

VER (Variable Energy Resource)
A renewable energy source, such as wind, solar, non-storage hydro generation or any other intermittent energy source, that is not carbon-emitting.

VFD (Variable Frequency Drive)
An AC motor technology that automatically adjusts a motor's speed to match output requirements to reduce energy use.

Virtual vertical
A market concept that advocates for development of power industry verticalization, characterized as a utility controlling some generation, some transmission, and VSA (Virtual Service Aggregator) functions, as a business model for successful Smart Grid operations.

Virtual water
See embodied water.

VLPGO (Very Large Power Grid Operators)
A global industry association of electric grid operators focused on identifying and developing emerging technologies such as HVDC and energy storage integration, defining solutions and best practices, and sharing knowledge. (http://www.vlpgo.org/)

VNM (Virtual Net Metering)
A tariff designed to offer the benefits of on-site solar generation in low-income multi-family housing without requiring the generation components to physically connect to each billing meter. Each meter in a multi-family housing unit shares in the net generation of electricity, with credits applying to monthly utility bills. One generation facility may therefore run multiple meters "backwards", or sell electricity back to the grid.

Volt (V)
The value of the difference, or voltage, across a conductor when a current of 1 ampere dissipates 1 watt of power in the conductor. Using a water analogy, the volt is the measure of water pressure, the amp measures the water's flow rate (current), and the ohm defines the pipe size (resistance).

Voltage collapse
When a utility's electric system does not have sufficient reactive power to maintain voltage stability. Total voltage collapse is a blackout condition. A partial voltage collapse is associated with voltage instability and affects only a portion of the utility's grid.

Voltage optimization
A practice involving capacitor banks, voltage regulators, distributed generating units, static VAR compensators, and other devices to maintain acceptable voltages at all points in the distribution grid and operate as efficiently as possible to save energy. It may have implications for demand response planning.

Voltage sag
An abrupt power disturbance generally caused at the start of large loads or during momentary overloads on the grid. It is the most common power disturbance impacting power quality.

Voltage stability
The ability of a power system to continuously maintain appropriate voltages within the system in normal conditions and during disruptions.

Voltage swell
An abrupt power disturbance generally caused by a reduction in load on a circuit with a malfunctioning voltage regulator.

Volttron™
This is a platform developed by the Pacific Northwest National Laboratory for use in the electric power system. It supports decentralized, cooperative decision-making amongst a wide range of distributed energy resources. Its primary function is to manage distributed generation, demand response, and plug-in electric vehicle grid assets. It was developed to support a DOE project called the Future Power Grid Initiative.

Volttron™ Lite
An open source agent execution platform that was developed as part of the DOE Transactional Network Project. It connects devices such as HVACs and power meters to applications in the platform, cloud, and signals from the power grid.

VPN (Virtual Private Network)
A network that uses security procedures and tunneling protocols to turn public networks like the Internet into the equivalent of a private network that offers secure access for remote offices or individual users back to corporate networks. Encryption technologies create tunnels that discourage access by unauthorized users. VPNs offer the benefits of true private or dedicated networks, but at a much lower cost.

VPP (Virtual Power Plant)
A collection of DR (demand response) programs and/or DER (distributed energy resources) that can be managed to equate to power plant capacity.

VRB (Vanadium Redox Battery)
A type of flow battery, the term redox describes electrochemical reactions in which energy is stored in two solutions. The VRB capacity ranges from 1kW up to several megawatts, and it has high-energy density and long life, which make it an energy storage possibility for utility load leveling, storage in renewable energy systems, and uninterruptable power supplies.

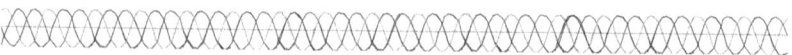

VRF (Violation Risk Factor)
An element used by NERC (North America Electric Reliability Corporation) to determine sanctions when any one of 83 Reliability Standards for the high voltage transmission system is violated. The factors are classified as high, medium, or lower risk. FERC (Federal Energy Regulatory Commission) approved 700 VRFs.

VSA (Virtual Service Aggregators)
A business model in which an entity dispatches and controls distributed sources of energy plus energy storage devices, and manages demand response and smart EV (Electric Vehicle) charging services for residential and/or C&I (Commercial and Industrial) customers. This entity aggregates generation and/or demand and manages relationships between utilities and end users.

VSC (Voltage Source Converter)
Technology used for reactive power compensation in high voltage DC environments.

VSL (Violation Severity Level)
A FERC (Federal Energy Regulatory Commission) term identifying the degree of violation (lower, moderate, high or severe) from a Critical Infrastructure Protection (CIP) or non-CIP reliability standard. This ranking helps define the financial penalty for the violation of reliability requirements.

VTN (Virtual Top Node)
A term found in the Energy Interoperation specification that describes an asset or device controller. It provides aggregated management of electric resources and capabilities within its domain such as aggregating demand response functions. Availability of resources aggregated by a VTN appears in the form of a single resource similar to that of a VPP (Virtual Power Plant).

Vulnerability
A cyber security gap or weakness that can lead to potential threats to confidentiality, availability, or integrity of data, devices, systems, grids, or networks.

Vulnerability assessment
Required of all water utilities serving more than 3,300 people to identify system vulnerabilities to terrorist attack or other intentional acts. The assessment includes system elements for transport, collection, pretreatment, treatment, storage and distribution; computer systems; and systems for operation and maintenance.

VVO (Volt-VAR Optimization)
Real-time applications that automatically coordinate voltage control and VAR losses across a distribution grid using advanced algorithms to model optimal load flows. Also known as advanced Volt-VAR control.

W

W (watt)
The basic unit that measures electric power, named after James Watt.

W3C (World Wide Web Consortium)
A global industry group that develops Web standards and guidelines. Since 1994, the W3C has published more than 110 such standards, called W3C Recommendations. It is also a forum for information, commerce, communication, and collective understanding. (http://www.w3.org/)

W3C EXI (Efficient XML Interchange)
This is W3C's an alternate binary encoding for XML. It is a data storage and transfer language very similar to XML but is more compact and is more efficiently exchanged and processed.

W3C SOAP (Simple Object Access Protocol)
This is World Wide Web Consortium standard for structured web services communication over HTTP. It is written in XML, and can therefore be used for Smart Grid device communication when needed.

W3C WSDL (Web Service Definition Language)
This is a World Wide Web Consortium standard for defining web services interactions. It allows descriptions of endpoints and their messages regardless of the original message format or network protocol. WSDL can be used for interoperability between Smart Grid devices.

W3C XML (eXtensible Markup Language)
This is a core standard for structuring and exchanging data and information. XML is widely used for the communication and storage of data via the Internet, and is also used for communication between devices in a Smart Grid.

W3C XSD (XML Schema Definitions)
A standard schema developed by the W3C for defining XML data instances, which are commonly found in WSDL (Web Service Definition Language), Web Services, and other XML applications.

WAC (Wide Area Control)
Realtime control strategy conducted on interconnected transmission networks by utilities and REs (Regional Entities) to detect and mitigate grid instabilities. It uses WASA (Wide Area Situational Awareness) to identify the emergence of grid instabilities within a geographic region that may correspond to an ISO (Independent System Operator) or RTO (Regional Transmission Organization) or a utility's footprint.

WADE (World Alliance for Decentralized Energy)
A global association that works to accelerate the development of co-generation, onsite power, and decentralized renewable energy systems. Members include energy associations, energy producers, generation equipment manufacturers, and public and private institutions. (http://www.localpower.org/)

WAMACS (Wide Area Measurement and Control Systems)
Systems that use PMUs to measure and transmit real-time status about power system equipment states on a wide geographic basis and deliver proactive, preventive, and emergency control and automated restoration.

WAMPAC (Wide Area Monitoring, Protection, and Control)
Systems that use synchrophasors and PMUs and advanced analytics to sense, visualize and control assets and operations in transmission grids.

WAMS (Wide Area Monitoring System)
Software and hardware that deliver a dynamic, geographic view of generation and transmission conditions through GPS-synchronized PMUs (Phasor Measurement Units). Functions include monitoring of thermal and

voltage limits and alarms that help utilities identify problems and respond to contain and resolve issues on the grid. It provides NERC (North American Electric Reliability Corporation) with alerts that the balance between generation and load has deviated significantly from scheduled values in specific control areas, and provides the location and amount of this deviation. The WAMS is also used by NERC Reliability Subcommittees and Working Groups for reliability performance tracking, analysis, and resource inadequacy post-assessment.

WAN (Wide Area Network)
A wired or wireless communications network that is geographically distributed. WANs can connect multiple LANs and other network types. Examples include public networks, such as telephone, wireless phone, wireless data, and the Internet.

WAP (Wide Area Protection)
Strategies to reduce the frequency and severity of grid disturbances that could result in wide area blackouts, including proactive and preventive grid upgrades, operator training, and sophisticated tools that enable realtime decisions regarding grid stability and performance. RAS (Remedial Action Scheme) and SIPS (System Integrity Protection Schemes) are two examples of WAP.

WAP (Wireless Application Protocol)
A set of standards that allow wireless mobile devices like phones to browse content from Web pages with specific coding.

WAPA (Western Area Power Administration)
A federal agency within DOE (Department of Energy) that markets and delivers hydroelectric power and related services within a 15-state region of the central and western USA (http://www.wapa.gov/)

War driving
Mobile search by driving around in vehicles for wireless networks that can be accessed for malicious intent.

WASA (Wide Area Situational Awareness)
Extensive realtime monitoring conducted on interconnected transmission grids by utilities and REs (Regional Entities) to maintain adequate load margins and avoid critical outages. Dynamic information is provided through a variety of devices, including PMUs (Phasor Measurement Units) across high-speed networks.

Wastewater
Water returned from residential, commercial, and industrial facilities that needs treatment prior to discharge back into the environment. It includes gray water, black water or water runoff from rainfall.

Wastewater treatment
Processes at a wastewater treatment plant that take contaminated water and clean it for additional use or discharge into the environment.

Water
A compound of hydrogen and oxygen that exists in solid, liquid, and gaseous forms. A key component for life, it has multiple uses, including electricity generation and storage.

Water collection point
A component of a water supply chain. Untreated water is stored above or below ground in lakes, rivers or aquifers.

Water distribution grid
See Water supply system.

Water Education Foundation
A nonprofit organization that promotes public knowledge and resolution of water resource issues through facilitation, education and outreach. (http://www.watereducation.org/)

Water footprint
A calculation of the total amount of water that is directly or indirectly consumed in the lifecycle of a product or service.

Water Footprint Network
A global nonprofit promoting a transition to sustainable use of fresh water resources through common consumption metrics, education and policy. (http://www.waterfootprint.org/)

Water mains
The grid of water distribution facilities that is located close to end users.

Water neutral
A process, building, or system that does not increase total water demand through water efficiency technologies and practices.

Water reclamation
The practice of reusing gray water for landscape irrigation rather than piping into waste water systems.

Water Resources Research Act of 1964
Federal legislation that created a water resources research and technology institute or center in each state. The objectives are to facilitate research on water problems; encourage entry into water resources fields; help train future water scientists and engineers; and distribute sponsored research to water managers and the public. The program is administered by the USGS and is also known as the "Water Resources Research Act Program" in the Department of the Interior.

Water source
Water originating from groundwater or aquifers, surface water such as rivers or lakes, conservation activities, and oceans via desalination. It is typically treated prior to distribution.

Water storage facility
A component of a water supply system that includes reservoirs, water tanks, towers, cisterns and pressure vessels.

Water supply system
A system of engineered hydraulic and hydrologic components that supply water. It typically includes a water source, collection point, purification facilities, distribution facilities, and treatment facilities. Also known as a water supply network or a water distribution grid.

Water table
The upper or top surface of groundwater. The soil below a water table is completely saturated with water.

Water treatment plant
Facilities that treat water before it is sent to the consumer and after the consumer has used it. The process of water purification usually occurs close to the end user to reduce pumping costs and the chances of subsequent contamination. Most treated water is delivered to the user via underground pipe systems.

WateReuse Association
A nonprofit organization promoting the uses of high-quality, locally produced and sustainable water sources through advocacy, education and outreach, research, and membership.
(http://www.watereuse.org/association)

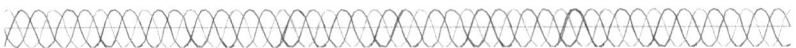

WateReuse Research Foundation
A nonprofit organization for the water and wastewater community to advance the science of water reuse, recycling, reclamation, and desalination through applied research and education. (http://www.watereuse.org/foundation/about)

WaterISAC (Water Information Sharing and Analysis Center)
An early warning system of potential threats to water and wastewater systems for utility operators. It also contains information about water system security topics obtained from federal intelligence agencies, law enforcement, public health and environmental agencies, as well as from utility security incident reports. Analysts provide both physical and cyber security information to the water sector to improve their ability to better develop and implement safety measures and response plans. (http://www.epa.gov/safewater/watersecurity/pubs/waterISACFactSheet.pdf)

WaterSense®
A program created by the US EPA to encourage greater water efficiency and enhance the market for water-efficient products, programs, and practices. It also establishes and standardizes rigorous certification criteria that ensure product efficiency, performance, and quality and identifies it with a labeling program. (http://www.epa.gov/WaterSense/)

Watershed
A region draining into a river, river system, or other body of water, it is the first link in a water supply chain.

WaterWiser®
A clearinghouse established by the AWWA to assist both water conservation professionals and the consumer in more efficient water use practices. It covers demand management, water conservation and efficiency in water use.

Wavenis® Open Standard Alliance
An international nonprofit SDO (Standard Development Organization) that promotes the use of Wavenis, a two-way wireless platform for ultra-low-power and long-range devices. (http://www.wavenis-osa.org/)

WBEM (Web-Based Enterprise Management)
A set of management and Internet technologies developed to unify the management of distributed computing and cloud based systems. It enables industrial integration of standards based management tools in a cross-platform fashion.

WCA (Wireless Communications Alliance)
A nonprofit business league dedicated to providing education and networking for companies, organizations, and individuals involved with wireless technologies. It includes a Smart Grid SIG (Special Interest Group). (http://www.wca.org/)

WCEE (Women's Council on Energy and Environment)
A nonprofit association that provides non-partisan and policy-neutral forums and education programs for people working in energy and environment. (http://www.wcee.org/)

WCI (Western Climate Initiative)
A GHG (Greenhouse Gas) reduction program adopted by seven states and four Canadian provinces (Arizona, California, Montana, New Mexico, Oregon, Utah, Washington, British Columbia, Manitoba, Ontario, and Quebec). It is a cap-and-trade program that goes beyond CO2 and regulates all fuels, including those used in transportation, industrial, and residential sectors. Several other USA states, Mexican states, and Canadian provinces are observers. (http://www.westernclimateinitiative.org/)

WEC (World Energy Council)
The association that represents the global energy industry. Its mission is to promote the sustainable supply and use of energy for the greatest benefit of all. The organization has official consultative status with the United Nations. It covers all types of energy, including coal, oil, natural gas, nuclear, hydro, and renewables. (http://www.worldenergy.org/)

WECC (Western Electricity Coordinating Council)
The organization responsible for coordinating and promoting electric system reliability and supporting efficient competitive power markets. It is the largest and most diverse of the eight NERC (North American Electric Reliability Corporations) Regional Entities or reliability regions. The WECC's territory includes the provinces of Alberta and British Columbia, the northern portion of Baja California, Mexico, and all or portions of Washington, Oregon, California, Arizona, Nevada, Idaho, Utah, Colorado, Montana, New Mexico, South Dakota, and Wyoming. The WECC region encompasses nearly 1.8 million square miles. It was formed in April 2002 by the merger of WSCC, Southwest Regional Transmission Association (SWRTA), and Western Regional Transmission Association (WRTA). AESO and CAISO perform market operations in this region. (http://www.wecc.biz/)

WEF (Water Environment Federation)
A not-for-profit association that provides technical education and training for water quality professionals focused on water treatment and restoration. (http://www.wef.org/)

Weightless
A proposed M2M standard that uses white space spectrum, low power, and reduced overhead to minimize transmission requirements for time and battery power. (http://www.weightless.org)

WEIL (Western Electric Industry Leaders)
A group that comprises CEOs and executive leaders from numerous entities in the western United States who are responsible for delivering electric energy to wholesale and retail customers. The goal is to ensure that renewable energy sources can be seamlessly integrated into the electric grid. Members include investor-owned utilities, municipalities, government agencies, and regional transmission operators, among others. (http://www.weilgroup.org/)

WERF (Water Environment Research Federation)
A nonprofit organization dedicated to scientific research on wastewater and stormwater issues. Members include wastewater treatment plants, stormwater utilities, regulatory agencies, industry, equipment companies, engineers, and environmental consultants. (http://www.werf.org)

West Coast Green Highway
A tri-state initiative creating a network of EV (electric vehicle) DC fast charging stations along Interstate 5 through the states of Washington, Oregon, and California. It will span 1,300 miles with public fast charging locations every 25-60 miles. Each location also includes Level 2 equipment to re-charge other EVs at slower charging rates too. (http://westcoastgreenhighway.com/electrichighways.htm)

Western Interconnection
The interconnected electrical systems within the WECC region, extending from Canada to Mexico.

Wet recirculating cooling systems
Part of a thermoelectric generation plant, systems that cool water using either cooling towers or cooling ponds.

Wh (Watthour)
A measure of energy, denoted as 1 Watt of power used for 1 hour.

Wheeling
The transmission of electricity generated by one party to another using the transmission system of a third party. The wheeler does not own, generate, or purchase the electricity being transported.

White space communication
The ability to leverage unused licensed spectrum, typically allocated for TV signals for other, unlicensed uses. It is contingent upon regulatory approval since it is licensed spectrum, but could help ensure the most efficient use of spectrum. It may have applications in rural broadband, smart city, and M2M communications.

Wholesale power
The production, delivery, and sale of electricity to the distribution utilities. This electricity is usually carried over a high-voltage, high-capacity transmission grid, which connects power plants to substations located near populated areas. Wholesale power is typically controlled by federal agencies.

WHRB (Waste Heat Recovery Boiler)
A type of heat exchanger that recovers energy from a hot exhaust stream. The waste heat is used to heat a boiler that generates steam, which can then be used for central heating or electricity generation via a turbine.

Wi-Fi (Wireless Fidelity)
An IEEE (Institute of Electrical and Electronics Engineers) standard 802.11 that refers to a family of specifications developed by the IEEE for wireless LAN technologies that use unlicensed radio spectrum. The term Wi-Fi initially described operations in the 2.4-GHz band, but the term has also been applied to unlicensed wireless devices operating in the 5-GHz band in

accordance with IEEE 802.11a. Wi-Fi technologies may also work in licensed spectrum. The FCC (Federal Communications Commission) does not require devices operating in unlicensed spectrum to meet the IEEE standards. The IEEE 802.11i standard addresses security issues with Wi-Fi.

Wi-Fi Alliance®
The global nonprofit Wi-Fi organization that created the Wi-Fi brand. It certifies interoperability of IEEE (Institute of Electrical and Electronics Engineers) 802.11 products and promotes them in all market segments. Members include companies that support the IEEE 802.11 family of standards. It recently announced an agreement to collaborate with the ZigBee® Alliance on wireless HANs (Home Area Networks) for Smart Grid applications. (http://www.wi-fi.com/)

Wi-SUN Alliance (Wireless Smart Utility Networks Alliance)
An industry alliance focused on certification and promotion of an interconnect standard for wireless Smart Grid devices, including smart meters and home appliances. It supports the emerging IEEE 802.15.4g standard for wireless communications of battery-powered equipment. (http://www.wi-sun.org)

Wideband
Information capacity or bandwidth that is between 64 kbps and 2 Mbps.

WiGig (Wireless Gigabit Alliance)
An industry alliance of vendors in the consumer electronics, computer, semiconductor and handheld industries with the mission to establish a global ecosystem of high-speed and easy-to-use wireless devices using unlicensed 60 GHz spectrum. (http://wirelessgigabitalliance.org/)

WiMAX (Worldwide Interoperability for Microwave Access)
Broadband wireless technology based on the IEEE (Institute of Electrical and Electronics Engineers) 802.16 family of standards. It provides high-

throughput broadband connections over long distances. WiMAX can be used for a number of applications, including last-mile broadband connections, hot spots (combined with WiFi), cellular backhaul, and high-speed enterprise connectivity without the need for line of sight to a base station. It operates in FCC-licensed and -unlicensed frequencies. Licensed WiMAX operates in the 10-to-66 GHz range; unlicensed WiMAX operates in the 2-to-11 GHz range.

WiMAX Forum®
A not-for-profit trade association of about 500 members that certifies and promotes the compatibility and interoperability of broadband wireless products based on the harmonized IEEE (Institute of Electrical and Electronics Engineers) 802.16/ETSI (European Telecommunications Standards Institute) HiperMAN standard. Their certified products are fully interoperable and support broadband fixed, portable, and mobile services. One of their working groups is developing specifications for fixed, nomadic, portable, and mobile WiMAX systems beyond the scope of IEEE 802.16, mostly based on inputs from a service provider working group. 802.16m is the next-generation standard beyond 802.16e-2005 and will be incorporated into the WiMAX System Release 2.0 in 2012. (http://www.wimaxforum.org/)

WINA (Wireless Industrial Networking Association)
A nonprofit industry coalition that provides information and educational outreach programs about the economic value and industrial applications for wireless technologies. (http://www.wina.org/)

Wind ramp
Infrequent, unforecasted, and large changes in wind production over short periods of time. Wind ramps consist of up-ramps (increases) or down-ramps (decreases) in wind production from thunderstorm activity.

Wind turbine
Equipment that harnesses wind energy and converts it to electricity. Turbines are either horizontal axis or vertical axis designs and consist of blades or rotors, shafts, drive trains, and supporting structures such as towers.

WINMEC (Wireless Internet for the Mobile Enterprise Consortium)
A UCLA-based university, industry, and government collaboration to advance wireless and mobile technologies and business research, and educate members on the state-of-the art in these industries. (http://winmec.ucla.edu/)

WInnF (Wireless Innovation Forum)
An industry association promoting Software Defined Radio (SDR), Cognitive Radio (CR) and Dynamic Spectrum Access (DSA) technologies for a broad range of applications. (http://www.wirelessinnovation.org/)

Wireless Communications
Wireless data communication devices typically use wireless transceiver modules that operate in the license-free Industrial Scientific and Medical (ISM) radio frequency bandwidths of 900 MHz and 2.4 GHz. Wireless communications in the 900-MHz bandwidth have up to twice the transmission range and better ability to penetrate than 2.4 GHz. 900 MHz is available only in North America, South America, Australia, New Zealand, and Israel. Wireless communications in the 2.4-GHz band have license-free communications in most of the world.

WirelessHART^{TM} (Wireless Highway Addressable Remote Transducer)
A wireless networking technology based on IEEE 802.15.4. The RF mesh network protocol operates in the 2.4 GHz ISM band, and is used in industrial process networks and FANs (field area networks). Also known as IEC 62591Ed. 1.0.

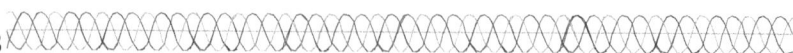

Wireless Power Consortium
An international association of companies focused on setting an international standard for compatible wireless charging stations for mobile devices powered by electricity. The wireless power transmission is based on the principle of magnetic induction.
(http://www.wirelesspowerconsortium.com/)

WIRES (Working Group for Investment in Reliable and Economic electric Systems)
A nonprofit group established to identify and advocate for policies that facilitate and promote needed investment in transmission on a national basis and among diverse groups within and outside the industry. Its members include electric transmission owners, ISOs (Independent System Operators), and utilities in the North American energy market.
(www.wiresgroup.com).

WISE (Water Infrastructure Security Enhancement)
An EPA initiative with grant funding to facilitate the development of guidance, training, and voluntary standards that cover the design of online contaminant monitoring systems and physical security enhancements of drinking water, wastewater, and stormwater infrastructure systems.

WISP (Western Interconnection Synchrophasor Program)
A synchrophasor measurement system deployed by WECC and partner entities to identify and analyze system vulnerabilities and evolving disturbances on the Western bulk electric system. The system improves the overall reliability of the Western grid and enables more integration of renewable energy sources.

Withdrawal
Water that is removed from underground sources or diverted from ground sources for use in industrial, commercial, agricultural, or residential applications.

WMC (The Watershed Management Council)
An international, nonprofit organization promoting watershed management through identification of research needs, legislative and policy advocacy, training, and public education. (http://watershed.org/)

World Water Council
An international body with the mission to promote awareness, build political commitment and trigger action on critical water issues at all levels and facilitate the efficient management and use of water in all its dimensions and on an environmentally sustainable basis (http://www.worldwatercouncil.org/)

Worm
A malicious software program that replicates itself across a network.

WOW (Wind On the Wires)
A coalition of wind developers, environmental organizations, wind energy experts, tribal representatives, clean energy advocates, and businesses focused on resolving technical, regulatory and educational issues to bring wind energy to market. (http://windonthewires.org/)

WPA2 (Wi-Fi Protected Access 2)
Another term for IEEE 802.11i, the approved, interoperable implementation secure wireless communications. It supersedes WPA and WEP (Wired Equivalent Privacy).

WPAN™ (Wireless Personal Area Network)
A wireless computer network that has a range of only a few yards or meters. Bluetooth and ZigBee are examples of technologies that support WPANs.

WPC (Wireless Power Consortium)
An industry association promoting wireless charging of devices using magnetic induction through a product interoperability standard. (http://www.wirelesspowerconsortium.com/)

WQA (The Water Quality Association)
A nonprofit, international trade association representing the water treatment industry, including businesses serving all sectors (residential, industrial, commercial and small communities). It provides professional education and public information and has a product-testing laboratory. (http://www.wqa.org/)

WREGIS (Western Renewable Energy Generation Information System)
An independent, renewable energy tracking system for the WECC (Western Electricity Coordinating Council) region. It tracks renewable energy generation from registered units and creates RECs (Renewable Energy Certificates) for this generation. These RECs can be used to verify compliance with state and provincial regulatory requirements and in voluntary market programs. One REC is equivalent to one MWh of renewable energy generation. (http://www.wregis.org/)

WREC/WREN (World Renewable Energy Congress/Network)
An international nonprofit organization that promotes environmentally safe and economically sustainable renewable energy sources. It aims to promote the communication and technical education of the renewable energy community and address the energy needs of both developing and developed countries. (http://www.wrenuk.co.uk/)

WRF (Water Research Foundation)
An international nonprofit organization comprised of water utilities, consulting firms and manufacturers that sponsors research into drinking water. Formerly known as the AWWA Research Foundation. (http://www.waterrf.org/Pages/WaterRFHome.aspx)

WS-MAN (Web Services Management)
This standard developed by DMTF (Distributed Management Task Force) now adopted as ISO/IEC 17963 is designed to provide a common, interoperable way for systems to access and exchange management information across an IT infrastructure. It is used as a network access protocol by many CIM (Common Information Model)-based management solutions.

WSDL (Web Services Description Language)
An XML-based language for describing Web services and how to access them. It includes interface information about all publicly available functions, data type for message requests and responses, transport protocol information, and address information.

WSN (Wireless Sensor Networks)
A wireless network of distributed monitoring devices or sensors that acquire and communicate information to a management center. Monitoring usually covers environmental or physical conditions, such as vibration, temperature, pressure, sound, or motion.

WTB (Wireless Telecommunications Bureau)
An office within the FCC (Federal Communications Commission) that handles cellular telephone, paging, personal communications services, and other commercial and private radio services programs, policies, and outreach initiatives except those involving public safety, satellite communications, or broadcasting. It is also responsible for implementing the competitive bidding authority for spectrum auctions, given to the Commission by the 1993 Omnibus Budget Reconciliation Act. (http://wireless.fcc.gov/)

WUCA (Water Utility Climate Alliance)
A group of ten of the nation's largest water and wastewater utilities that provide services to 43 million Americans. The goal of WUCA is to provide leadership and collaboration on climate change issues that will affect the United States' water infrastructure in the future.

X/Y/Z

X10

A home networking protocol based on PLC (Power Line Communications) that distributes communications signals using the electrical wiring.

XMPP (Extensible Messaging and Presence Protocol)
An open technology for real-time communication for instant messaging, network management, presence, lightweight middleware, generalized routing of XML data, among other applications. IETF's (Internet Engineering Task Force) XMPP Working Group formalized the core protocols and published them as RFCs (Request for Comments) 3920 and 3921.

XSF (XMPP Standards Foundation)
An independent, nonprofit standards-development organization whose primary mission is to define open protocols on top of the IETF's (Internet Engineering Task Force) XMPP (Extensible Messaging and Presence Protocol). It also provides information and support to the worldwide community of XMPP developers, service providers, and end users. (http://xmpp.org/xsf/)

ZBB (Zinc Bromide Battery)
A type of flow battery that may be suitable for utility-scale energy storage due to its high-energy density and low-cost materials. The capacity ranges from 5 to 20 megawatts. It is also known by the acronym ZnBr and as Zinc Bromine.

ZEBRA (Zero Energy Building Research Alliance)
A research and education project focused on residential energy efficiency. Members include the Schaad Companies, Tennessee Valley Authority (TVA), Oak Ridge National Laboratory (ORNL), BarberMcMurry Architects, and the DOE (Department of Energy). The ORNL's energy-efficient technologies are integrated into homes to test functional and cost effectiveness. (http://www.zebralliance.com/)

Zero day exploit
Any hacking activity that occurs the same day that any security vulnerability is discovered.

Zero net energy homes
The Long-term Energy Efficiency Strategic Plan recently adopted by the CPUC (California Public Utilities Commission) requires that all new residential construction provide "zero net energy" by 2020, with new commercial construction to comply by 2030. Any building's annual consumption of electricity or natural gas from utility suppliers must be offset by a combination of energy-efficient building features and distributed generation.

ZigBee®
A wireless mesh networking specification based on the IEEE (Institute of Electrical and Electronics Engineers) 802.15.4 standard to enable low-cost, low-power wireless sensor networks for HAN (Home Area Network) applications and other uses. It operates in unlicensed bands worldwide at the following frequencies: 2.400 to 2.484 GHz, 902 to 928 MHz, and 868.0 to 868.6 MHz. It supports data transmission rates of up to 250 Kbps with a range of 250 feet, and optionally supports encryption. It is slower than WiFi at 11 Mbps and Bluetooth at 1 Mbps, but uses less power. IETF standards will add native IP support to ZigBee, and the RF4CE (Radio Frequency for Consumer Electronics Consortium) will collaborate with the Zigbee Alliance on a specification for RF-based remote controls for home entertainment devices.

ZigBee® Alliance
An industry association that includes 300 electric utilities, meter manufacturers, and industrial automation and design companies promoting ZigBee for use in wireless networks as a sensing and control standard in consumer electronics, energy, home, commercial, and industrial areas. The HomePlug® Powerline Alliance, WiFi® Alliance, and EPRI (Electric Power

Research Institute) are working with this association to develop a common language, standard communication approach, and common set of certification procedures for HAN (Home Area Network) devices to use AMI (Advanced Metering Infrastructure) technology. (http://www.zigbee.org/)

ZigBee® IP 2.0
Another term for the Smart Energy Profile 2.0, an IP-based energy management solution that is MAC (Medium Access Control) and PHY (Physical) agnostic.

Z-wave
A wireless technology for HAN that uses low-power radio waves that easily pass through home interiors. Z-Wave is a two-way routing protocol for RF mesh networks. This protocol was developed by Zensys, a division of Sigma Designs, Inc.

Z-wave Alliance
A consortium of home controls manufacturers that use the Z-Wave HAN (Home Area Network) standard. They work to promote interoperability between systems and devices and cooperation between manufacturers on future products and services. (http://www.z-wavealliance.org/modules/start/)

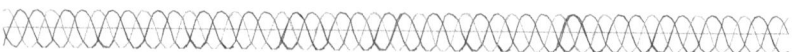

m/pod-product-compliance